Recent Results in Cancer Research

Fortschritte der Krebsforschung

Progrès dans les recherches sur le cancer

25

Edited by

*V. G. Allfrey, New York · M. Allgöwer, Basel · K. H. Bauer, Heidelberg
I. Berenblum, Rehovoth · F. Bergel, Jersey · J. Bernard, Paris · W. Bernhard, Villejuif · N. N. Blokhin, Moskva · H. E. Bock, Tübingen · P. Bucalossi, Milano · A. V. Chaklin, Moskva · M. Chorazy, Gliwice · G. J. Cunningham, Richmond · W. Dameshek †, Boston · M. Dargent, Lyon · G. Della Porta, Milano · P. Denoix, Villejuif · R. Dulbecco, La Jolla · H. Eagle, New York
R. Eker, Oslo · P. Grabar, Paris · H. Hamperl, Bonn · R. J. C. Harris, London
E. Hecker, Heidelberg · R. Herbeuval, Nancy · J. Higginson, Lyon
W. C. Hueper, Fort Myers · H. Isliker, Lausanne · D. A. Karnofsky †, New York · J. Kieler, København · G. Klein, Stockholm · H. Koprowski, Philadelphia · L. G. Koss, New York · G. Martz, Zürich · G. Mathé, Villejuif
O. Mühlbock, Amsterdam · W. Nakahara, Tokyo · V. R. Potter, Madison
A. B. Sabin, Rehovoth · L. Sachs, Rehovoth · E. A. Saxén, Helsinki
W. Szybalski, Madison · H. Tagnon, Bruxelles · R. M. Taylor, Toronto
A. Tissières, Genève · E. Uehlinger, Zürich · R. W. Wissler, Chicago
T. Yoshida, Tokyo*

Editor in chief
P. Rentchnick, Genève

Springer-Verlag New York · Heidelberg · Berlin 1970

Analogues of Nucleic Acid Components

Mechanisms of Action

By

P. Roy-Burman

With 41 Figures

Springer-Verlag New York · Heidelberg · Berlin 1970

P. Roy-Burman, Ph. D., Assistant Professor of Biochemistry,
University of Southern California, School of Medicine, Los Angeles, CA/USA

Sponsored by the Swiss League against Cancer

ISBN-13: 978-3-642-85578-8   e-ISBN-13: 978-3-642-85576-4
DOI: 10.1007/978-3-642-85576-4

*This monograph is dedicated to
The American Cancer Society*

Contents

Acknowledgments

The author wishes to express his sincere appreciation for critical reading of the manuscript and many helpful suggestions to Dr. DONALD W. VISSER, Professor of Biochemistry, and to Dr. RICHARD L. O'BRIEN, Assistant Professor of Pathology, University of Southern California School of Medicine. The author is also grateful to Dr. DANIEL LEVY, Assistant Professor of Biochemistry, and to Mr. RICHARD A. PASELK, Graduate Student of Biochemistry of this School, for their comments and suggestions on various sections of the monograph.

The author owes a special debt of gratitude to his wife, Dr. SUMITRA ROY-BURMAN for initially stimulating his interest in writing this monograph, and for her continual encouragement, patience, tolerance and invaluable help in preparing the manuscript. The author is indebted to Mr. CARLOS M. MEDINA for help with the references, and to Mrs. PENELOPE V. ACHILLES, Mrs. MARCIA E. LUTWEN, Miss MARY E. SIMONS, and Mrs. VARIAN L. HAGGLUND for their able secretarial assistance.

During the preparation of this monograph, the author has been supported by Grant T-478 from the American Cancer Society, and by Grant D-120 from the American Cancer Society, California Division.

Chapter 1

Introduction

The rationale for the design of structural analogues of a normal metabolite is that such compounds may interfere in the utilization or function of the metabolite. A compound which is effective in this respect may be called an antimetabolite. To be successful in chemotherapy of bacterial, viral, or tumor growth, an antimetabolite should adversely affect some vital metabolic reactions in the parasite or parasitic tissue without seriously endangering the host tissue. If a metabolic process of the offending growth is different from that of the host, it is likely that the metabolism or activity of a compound, structurally related to a metabolite involved in that process, will also be different in these cells. Such differences are useful for devising effective drugs with selective actions. Sulfanilamide, a structural analogue of *para*-aminobenzoic acid, interferes with the utilization of this metabolite in the synthesis of folic acid, an essential factor for growth. Bacteria synthesize their own folic acid and are incapable of utilizing exogenously available folic acid. However, the situation is exactly opposite in the animal host. That is, animal tissues cannot synthesize folic acid and are absolutely dependent upon exogenous sources. These differences in metabolism make possible the use of sulfanilamide as a selective inhibitor of growth. Other antibacterial or antiparasitic drugs, such as penicillin (BURCHALL, FERONE and HITCHINGS, 1965) and inhibitors of dihydrofolate reductase (HITCHINGS and BURCHALL, 1965; HITCHINGS, 1964; BURCHALL and HITCHINGS, 1965) have analogous desirable selective toxicity effects. Unfortunately, there are few qualitative differences in the metabolism of normal and neoplastic tissues. This fact has made the selective toxicity approach less applicable to cancer chemotherapy. However, it has become increasingly clear that there are quantitative differences which can be utilized as a rational approach to cancer chemotherapy. These considerations are based on the rate of metabolism and the state of cellular structures and are directly related to the rapid growth and multiplication of neoplastic cells.

Purines and pyrimidines are basic components of deoxyribonucleic acid (DNA) and ribonucleic acid (RNA). These heterocyclic moieties are also present in various coenzymes. In both normal and abnormal growths of a living system these universal components are involved in their specific expressions of cellular function and multiplication. This makes obvious the reasons and significance of the studies on the synthesis and biological evaluations of the analogues of purines and pyrimidines and their derivatives. It was about two decades ago when HITCHINGS and his co-workers (HITCHINGS et al., 1950; HITCHINGS et al., 1950 a) first began a systematic study of analogues of the purine and pyrimidine bases. Coincident with increasing knowledge of the important functions of nucleic acids in cell division and growth, it was quite natural to think of the possibility that nucleic acid synthesis may be preferentially inhibited in rapidly growing cells, like neoplastic tissues, by the use of base analogues. This has led to the preparation of thousands of such compounds and determination of their antitumor and other biological activities. The progress made in understanding the details of the modes of action has revealed that, in most instances, the inhibition produced by base or nucleoside analogues is due to one or more of the following mechanisms.

1. The inhibitor may mimic the normal metabolite and bind with the active site of an enzyme, thereby inhibiting the formation of the substrate-enzyme complex. This type of inhibitor enzyme combination may be completely reversible, partly reversible or in some cases irreversible, depending on the binding properties of the inhibitor to the active site. An analogue that is converted to various derivatives may, in an analogous manner, produce inhibition of several enzymes. Base or nucleoside analogues may be phosphorylated to nucleoside mono-, di- or triphosphates by nucleoside phosphorylase, base phosphoribosyltransferase and kinases, and these phosphorylated derivatives may be involved in inhibition of specific enzymes catalyzing similar reactions. Usually, conversion to nucleotide(s) is necessary for most purine, pyrimidine and nucleoside analogues to exert their inhibitory effects. Analogues of nucleoside triphosphates, when formed, may be expected to compete with specific natural nucleoside triphosphates in their utilization for DNA or RNA synthesis, resulting in inhibition of the formation of these macromolecules.

2. The analogues may be incorporated into nucleic acids from its nucleoside triphosphate level. When certain analogues are incorporated into DNA, the frequency of mutations may be increased. Incorporation into DNA in place of a normal constituent may be expected to interfere with or cause errors in the process of replication of DNA and also in the process of transcription where RNA molecules are produced on the DNA template. There are several types of RNA within a cell, namely, transfer RNA—responsible for transfer of activated amino acids for protein synthesis, ribosomal RNA—a structural unit of ribosomes, and messenger RNA which carries the code for the synthesis of specific proteins. Incorporation of analogue into RNA, thus, may inhibit or modify its function by interference with the reading of the genetic code. The net result is inhibition of protein synthesis or conceivably, the synthesis of abnormal polypeptides.

3. The analogue may interfere with cellular control mechanisms. A derivative of an inhibitor when accumulated in the cell may exert control by mimicking a normal controlling metabolite. One might, therefore, envisage a cellular growth inhibition by end product or feedback inhibition through allosteric effects or by repression mechanisms. An analogue which is metabolized to several different abnormal products may inhibit activity of various enzymes by feedback effects. Similarly, these abnormal products may repress production of enzymes. In addition to these negative phenomena, analogues may produce activation of production or catalytic function of enzymes by induction or positive feedback effects. However, examples of such effects with purine, pyrimidine, and nucleoside analogues are not common.

In recent years the mechanism of action of analogues has been studied extensively. With respect to purine and pyrimidine analogues it has become evident that they participate in many biochemical reactions and inhibit at multiple *loci*. In spite of the limited usefulness of most of these compounds as chemotherapeutic agents, studies concerning the mode of action of analogues have led to useful information and have provided new rationale for the design of analogues. An important by-product of these studies has been a better understanding of the biochemical, or more specifically enzymatic reactions in normal as well as abnormal growths.

Although an ideally effective drug for control of neoplastic growth is not yet in hand, some analogues are capable of prolonging life expectancy of cancer patients. A purine analog, 6-mercaptopurine, has been useful for treatment of childhood leukemia. A somewhat more limited success in the use of this compound has been achieved in the treatment of adult acute leukemia. FREI (1967) has reported that, at best, 10 to 15 per cent of adults with acute myelogenous leukemia achieve complete remission by 6-mercaptopurine treatment. 6-(Methylthio)purine ribonucleoside, thioguanine, and chloropurine have therapeutic effects similar to those of 6-mercaptopurine in human leukemias (ELION and HITCHINGS, 1965; MONTGOMERY, 1965). The pyrimidine analogues, 5-fluorouracil and 5-fluoro-2'-deoxyuridine have been used in the treatment of solid tumors. It has been reported by MONTGOMERY (1965) that about 20 to 30 per cent of the solid tumors treated respond, and the beneficial effects last from a few weeks to years. A recent report by KRAKOFF (1967) states that about 15 per cent of patients with breast cancer treated with these agents experience beneficial responses. According to REGELSON (1967) 5-fluorouracil is the best agent in the treatment of all gastrointestinal cancers. This analogue causes significant objective regression in about 25 per cent of patients with inoperable or recurrent metastatic gastrointestinal cancer (REGELSON, 1967). Clinical effects of azauridine or its triacetyl derivative has been observed in patients with psoriasis and mycosis fungoides (ZARUBA, KUTA and ELIS, 1963; CALABRESI, TURNER and LEFKOWITZ, 1964). CAREY and ELLISON (1965), and HOWARD, CEVIK and MURPHY (1966) observed that partial remissions of acute myelocytic leukemia or acute childhood leukemia can be obtained by the use of arabinosylcytosine. The purine nucleoside antibiotic, tubercidin, has shown preliminary clinical activity against some pancreatic tumors (HEIDELBERGER, 1967).

Some analogues which were designed for potential use in cancer chemotherapy have been shown to be useful for other chemotherapeutic purposes. 5-Iododeoxyuridine has been successfully used in the treatment of herpes keratitis, an eye disease caused by herpes simplex virus infection of the cornea. This application of iododeoxyuridine indicates that viral diseases may be controlled by the use of analogues and directs attention to the importance of the investigations in this field. Puromycin, a nucleoside antibiotic, has been used with partial success in infections with *T-gambiense*, the protozoa causing sleeping sickness (HAWKING, 1963). Allopurinol, 4-hydroxypyrazolo (3,4-d) pyrimidine, has been an effective compound for the treatment of gout. This purine analogue is capable of lowering serum and urinary uric acid concentration in patients with hyperuricemia and preventing uric acid stones in patients with hyperuricosuria (ELION, KOVENSKY, and HITCHINGS, 1966; ANDERSON et al., 1967; RUNDLES, METZ and SILBERMAN, 1966; WYNGAARDEN, 1966). Allopurinol is an exceptionally well-tolerated drug which has no apparent adverse effect on renal or hepatic function (ANDERSON et al., 1967; RUNDLES, METZ and SILBERMAN, 1966). The drug has applications in the purpose for which it was originally developed. It reduces the therapeutic dose of 6-mercaptopurine required in leukemic patients by inhibiting the conversion of 6-mercaptopurine to 6-thiouric acid (RUNDLES, METZ and SILBERMAN, 1966; ELION et al., 1963 a; RUNDLES et al., 1964; RUNDLES et al., 1963). By itself, the compound has little or no effect upon experimental tumors (WHITE, 1959; SHAW et al., 1960).

Pyrimidine and purine analogues have been useful in many other ways, in addition to their clinical usefulness. They are excellent tools for the study of many enzymatic reactions at the molecular level. Other applications are elucidation of the chemical basis of mutation, revelation of intermediary biochemical reactions through the imposition of specific metabolic blocks, and interpretation of the nature of base-pairing among nucleic acid components and polynucleotides. Discussion of the use of pyrimidine and purine analogues in the study of the metabolism of nucleic acids and their components in both normal and abnormal cell growths is beyond the scope of this review. However, it may be appropriate to cite some of these applications to illustrate the significance of the contributions of these analogues in the understanding of biochemical phenomena.

1. BESSMAN et al. (1958) showed that several pyrimidine and purine analogues are incorporated enzymatically into DNA from their deoxyribonucleoside triphosphate levels by DNA polymerase. 5-Bromodeoxyuridylic acid is incorporated into DNA specifically in place of deoxythymidylic acid and 5-methyl- or 5-bromodeoxycytidylic acid in place of deoxycytidylic acid. The specific replacement of the natural base by the analogues provided additional support for the base-pairing relationships in the double helix structure of DNA proposed by WATSON and CRICK (1953). KAHAN and HURWITZ (1962) observed that in the synthesis of RNA from ribonucleoside triphosphates by DNA-dependent RNA polymerase, 5-fluoruridylic acid and 5-bromouridylic acid are incorporated into RNA specifically in place of uridylic acid; 6-azaguanylic acid specifically replaces guanylic acid and 5-bromocytidylic acid replaces cytidylic acid. The analogue, 5-hydroxyuridylic acid specifically replaces uridylic acid in the same reaction (ROY-BURMAN, ROY-BURMAN and VISSER, 1966). Similar studies were carried out with ribonucleoside triphosphates containing base analogues in the viral RNA polymerase reactions and similar results were obtained (SHAPIRO and AUGUST, 1965). Thus, a fundamental similarity exists in the hydrogen bonding mechanisms by which DNA acts as a template for the formation of either complementary DNA or complementary RNA or by which a viral RNA makes its own complementary RNA.

2. GOLDBERG, DAHL and PARKS (1963) examined the effects of alterations in the pyrimidine portion of UDP-glucose upon its interaction with UDP-glucose dehydrogenase. Their studies with 5-fluoro-UDP-glucose and 6-aza-UDP-glucose revealed that the ionic state of the pyrimidine moiety has a determining effect on the binding of substrate to enzyme and that the undissociated form of the substrate is the effective species for interaction with the enzyme. Similar ionization effects were observed by ROY-BURMAN, ROY-BURMAN and VISSER (1968) with 5-hydroxy-UDP-glucose and 5,6-dihydro-UDP-glucose on the same enzyme and by KAHAN and HURWITZ (1962) in their studies of the synthesis of RNA from 5-halogenated uridine triphosphates. As pointed out by GOLDBERG, DAHL and PARKS (1963) these findings may be significant in chemotherapy since they suggest that uridine analogues, which have lower pKa values than uridine, may be more effective in cells which have an acidic environment. Thus, it is possible that the greater inhibitory effects of certain purine and pyrimidine analogues on cancer cells than on normal cells may be explained by the fact that tumor cells have a lower pH than normal cells from which they are derived.

3. The nucleoside analogues have been used in a very interesting manner to study the replication of DNA in human cells (ERIKSON and SZYBALSKI, 1963) and in bean roots (HAUT and TAYLOR, 1967). Bean roots were grown in solutions containing bromodeoxyuridine and also fluorodeoxyuridine. The latter analogue was used to inhibit the *in vivo* formation of deoxythymidylic acid by its specific inhibition of the enzyme deoxythymidylate synthetase and thereby facilitating the utilization of the former analogue for DNA synthesis. After various intervals of time DNA containing bromodeoxyuridylic acid in place of deoxythymidylic acid was isolated and analyzed by centrifugation in a density gradient. The atomic weight of bromine is much heavier than the molecular weight of the $-CH_3$ group. Accordingly, it was observed that the DNA of hybrid density increased in amount during the first replication cycle. After some cells had entered the second replication cycle, fully substituted DNA appeared as predicted on the basis of the mechanism of semiconservative replication. The same property of the increased density of the bromine-containing DNA permitted the recent accomplishment of the *in vitro* synthesis of biologically active DNA (GOULIAN, KORNBERG and SINSHEIMER, 1967). These authors have been able to produce infectious DNA of ΦX 174 virus in test tubes using the procedures as summarized in Fig. 1.

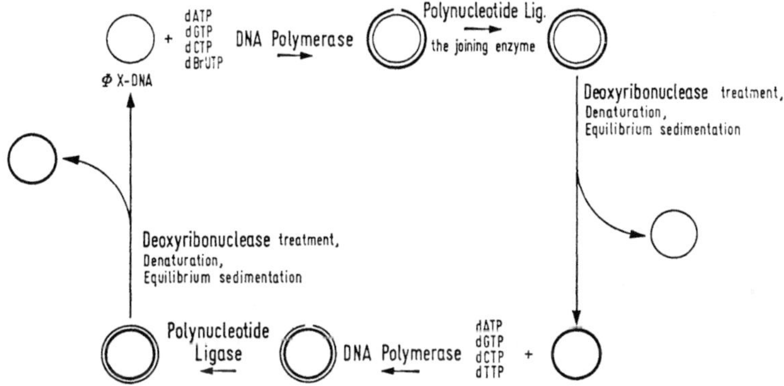

Fig. 1. Synthesis of infective viral DNA. (After GOULIAN, KORNBERG and SINSHEIMER, 1967)

The genome of ΦX virus is a single stranded covalently closed circular DNA. This genome, separated from the protein shell that covers it in normal, intact virus, is able to infect specially prepared forms of *Escherichia coli*, known as spheroplasts. Inside the cell, a second strand is synthesized which is the complement of the genome. The new structure is a double-stranded, base-paired form containing both the strands in circular forms. It was necessary to separate the DNA circles from this type of structures in order to determine the stages of replication and to physically distinguish the synthesized viral DNA molecules from the parent molecules. This was made possible by the use of the base analogue, bromouracil. In the mixture of the deoxyribonucleoside triphosphates used for polymerization the natural substrate dTTP was replaced by 5-bromodeoxyuridine triphosphate (dBUTP). The substitution of the methyl group in deoxythymidylic acid by the bromine atom permits a normal polymerase reaction and does not alter the infectivity of ΦX-DNA. Centrifugation in a density gradient resulted in the separation of the heavy strand of DNA containing bromodeoxyuridylic acid (heavy line) from the normal light strand of the ΦX-DNA (thin line), as illustrated in Fig. 1. These bromouracil containing DNA circles, which are obviously new complementary DNA molecules, were used to produce more DNA molecules in the presence of all normal substrates, dATP, dGTP, dCTP, and dTTP. These DNA circles containing normal bases, upon separation, were found to be infectious and chemically identical to those of ΦX-DNA. Thus, for the first time it was clearly shown that *in vitro* synthesis of biologically active DNA is possible.

There has been a wide interest recently in the study of antimetabolites as potential immunosuppressive drugs. This new field was opened by the work of SCHWARTZ, STACK and DAMESHEK (1958) who demonstrated that the administration of 6-mercaptopurine to rabbits which were exposed to a foreign antigen, bovine serum albumin, prevented antibody formation and made the animals unresponsive to later doses of the same antigen. The antimetabolites which are capable of inhibiting the immune response are currently being used to treat autoimmune diseases or to suppress the body's natural tendency to reject foreign tissue of transplantation. The subject has been reviewed by HITCHINGS and ELION (1963), and ELION and HITCHINGS (1965). Azathioprine (Imuran), 6-mercaptopurine, methylthiopurine ribonucleoside, and thioguanine, are purine analogues which have been used to suppress homograft rejection in experimental animals. Immunosuppression in human patients has been achieved primarily by the use of azathioprine along with steroids, azaserine, actinomycin, and X-irradiations (ELION and HITCHINGS, 1965). The World's first human heart transplantation was made in South Africa in December, 1967. To overcome the body's natural tendency to reject the implanted heart, the patient was given azathioprine, prednisone and cobalt radiation treatments. The patient died from pneumonia after 18 days. This infection was possibly acquired because the drugs and radiation drastically reduced his ability to fight germs as well as a foreign heart. Although some promising advances are being made in this direction, researchers have not yet found a selective immunosuppressive agent or a method of application that can effectively discriminate between these desirable and undesirable actions. As a consequence, infection is a predominant cause of death in persons who have received transplanted kidneys or other organs and immunosuppressive drugs or radiation. The mechanism of chemical suppression of immune response is also not known with any degree of certainty at this time. The possibility of a selective destruction of a small clone of cells, in which the rate of multiplication has increased in response to an antigenic stimulus, or a possible metabolic block of some informational processes by inhibition of nucleic acid biosynthesis or by incorporation of the base analogues into nucleic acids, has been postulated (HITCHINGS and ELION, 1963). It is realized that the base analogues would be very useful tools for the understanding of this phenomena and that this information in turn would be useful for clinical application. However, the existing problems in obtaining a pure culture of the cell line concerned with a specific immune response must be overcome before such studies can be carried out.

Analogues could be applicable in the studies of the newly developing field of comparative enzymology. It has been documented that the enzymes performing the same function may vary considerably in the nature of their reactivity and conformations from organism to organism. Most of these differences are detected by highly sensitive immunological techniques, by the selective effects of chemotherapeutic agents with a known *locus* of action, or by chemical and physical properties. Interest in the use of metabolic inhibitors for these studies arose from the findings of HITCHINGS et al. (1952) who showed that minor changes in the structure of 2,4-diaminopyrimidines or other derivatives could make them highly specific toward one organism or another in their inhibition of the utilization of folic acid. HITCHINGS (1967) has recently pointed out that "metabolic specificity may be the resultant of a series of molecular sieves of different degrees and kinds of specificities. Considerable latitude in the design of the individual sieves may be permissible as long as the end result is the same. When the system is dealing only with normal metabolites the variations are not apparent, but a foreign substance, a drug or an antimetabolite, may produce quite individualistic results. The interpretation and documentation of these at the molecular level have important implications not only for chemotherapy and comparative biochemistry and pharmacology, but also for an understanding of evolution in terms of molecular events."

It is apparent, from this brief and necessarily superficial introduction of the subject, that the analogues of purines and pyrimidines are biochemically interesting compounds and some of them are clinically useful. But at this point we should also consider the limitations which are encountered in their uses. Two major criteria which limit the therapeutic effectiveness of these analogues are, 1. a lack of tissue selectivity; 2. the development of cell lines resistant to their inhibitory effects.

For a specific destruction of an abnormal growth it is desirable that an analogue should be effective in exerting its inhibitory effect on that growth without causing interference with the normal growth. As discussed earlier, the rationale is that they are likely to exert their adverse effect on the faster metabolism in rapidly proliferative tissues than in normal tissues. In other words, if these compounds are visualized as inhibitors of cell division, it might be predicted that they would preferentially inhibit cell division of a rapidly dividing clone of cells. It is possible that the existence of this phenomenon is responsible for the partial success achieved in the application of analogues as antitumor agents. Unfortunately, for the same reason, rapidly proliferating normal tissues, especially intestinal mucosa, bone marrow, and embryos, are affected by the analogues. The ultimate goal is to overcome these toxic manifestations by developing more selective inhibitors through the design of new analogues. It is also realized that such efforts are dependent upon a greater insight into the biochemistry of both normal and abnormal growth.

The emergence of resistance is believed to result from a genetic mutation followed by selection of the resistant cells that have a proliferative advantage in the presence of the inhibitor. The theoretical possibilities involved in the modification of a drug-sensitive cell line to a drug-resistant one have been explored using a number of antimetabolites. These investigations have shown that during the influence of the drug, certain changes in the cellular metabolism occur to escape from the inhibitory effects of the drug. These alterations include decreased conversion of the inhibitor to an active form, increased degradation of the inhibitor to an inactive compound, increased synthesis of the inhibited enzyme, modification of an enzyme or enzyme system directly involved in the anabolism of the inhibitor, decreased permeability of resistant cells to an inhibitor, and possible emergence of alternate metabolic pathway which bypass the metabolite (BROCKMAN, 1963; HUTCHISON, 1965). The present knowledge regarding the mechanism of resistance formation will be included in the discussion of the mechanisms of actions of individual analogues.

In addition to these two main obstacles encountered in the clinical use of analogues another difficulty exists which is not as acute as the former two. The problem is that most of the purine and pyrimidine analogues are extensively catabolized. It is known that, usually purine, pyrimidine or nucleoside analogues must be converted to their corresponding nucleotide derivatives to exert their inhibitory effects. Catabolic reactions compete with these reactions, thus destroying their potency. For example, 6-mercaptopurine is catabolized irreversibly to thiouric acid and arabinosylcytosine is converted to arabinosyluracil. These products are ineffective analogues and are excreted. These undesirable side reactions may in some cases be eliminated by appropriate analogue design. For example, the conversion of 6-mercaptopurine to thiouric acid may be limited by simultaneous administration of 4-hydroxypyrazolo (3,4-d) pyrimidine (allopurinol) which inhibits xanthine oxidase. Under these conditions, smaller doses of 6-mercaptopurine are effective and a better tumor therapeutic index is achieved (FREI, 1967a).

The present discussion will consider aspects of the metabolism of purine and pyrimidine analogues and the biochemical mechanisms by which they inhibit metabolic processes. The enormous amount of work done in this area precludes a comprehensive review. Consequently, the present discussion will be limited mainly to the analogues which are useful in controlling viral or tumor growth or to the analogues which have been extensively studied in a wide variety of biochemical reactions. Moreover, the author's intent is to acquaint the reader with the over-all concept of this rapidly developing field, not to present an exhaustive review of the literature. Various aspects of the mechanism of action of base or nucleoside analogues have been discussed in the excellent reviews by BROCKMAN and ANDERSON (1963), ELION and HITCHINGS (1965), HEIDELBERGER (1967), COHEN (1966), and HENDERSON (1965). The naturally occurring nucleoside antibiotics may be considered as nucleoside analogues. The subject concerning their mechanism of action has been included in the review by FOX, WATANABE and BLOCH (1966), and in the first volume of "Antibiotics" edited by GOTTLIEB and SHAW (1967).

A note of caution should be inserted at this point. It should be understood that the detailed mechanisms considered in the following pages are those which have been studied. Additional inhibitory effects, which are as yet unknown, may have equal or even greater importance in living cells. The results under consideration have been obtained from both *in vivo* and *in vitro* experiments with bacterial, viral, animal and tumor systems. In addition to the obvious point that results from *in vitro* systems may not always be the same as in *in vivo* situations, the question exists regarding the species specificity of the various enzymes inhibited by an antimetabolite. It is also reasonable to believe that expressions of the inhibitory effects of an analogue are likely to be determined by the intactness of a cell as well as by its normal environmental conditions. With these reservations in mind, the identification of the major metabolic sites where the inhibitor or its metabolic products interfere with the utilization or function of normal substrates are discussed. These collected pieces of information derived from a wide spectrum of studies serve as a framework for the future approaches to the development of the field.

Chapter 2

Purines

A. Metabolism of Purines

As a basis for reviewing biochemical mechanisms of the action of purine analogues, the normal metabolic reactions of purines should be understood. For this purpose, a brief outline of purine metabolism is presented here. Various aspects of the subject have been considered by BUCHANAN and HARTMAN (1959), FLAKS and LUKENS (1963), SUGINO (1965), LARSSON and REICHARD (1967), and BLAKLEY and VITOLS (1968). These articles should be consulted in pursuing the subject further.

The following pathway for *de novo* purine biosynthesis (Fig. 2) appears to be operative in all organisms so far studied. The components of the purine ring are derived from formate, carbon dioxide, amide nitrogen of glutamine, amino group of aspartic acid, and glycine, and the ring is completed after conversion to a phosphorylated ribosyl derivative. Inosine 5'-monophosphate (IMP) is the parent purine compound formed in the *de novo* pathway and this is utilized for the synthesis of adenine and guanine nucleotides as outlined in Fig. 3. It should be noted that the reductase of *Escherichia coli* or animal tissues reduces ribonucleoside diphosphates to deoxyribonucleoside diphosphates, whereas the enzyme from *Lactobacillus leichmannii* shows requirement for ribonucleoside triphosphates as substrates. In Fig. 3 and where ever appropriate the nucleoside diphosphates are shown as substrates for reduction since most of the studies with purine and pyrimidine analogues which are included in the present discussion were carried out with *Escherichia coli* or animal systems.

In addition to the *de novo* pathways, there are *salvage pathways* or pathways that utilize intact bases which arise from prior cleavage of nucleotides (Figs. 4 and 5). Purine bases are converted to the corresponding ribonucleotides by reacting with 5-phosphoribosyl-1-pyrophosphate (PRPP). The reactions catalyzed by purine phosphoribosyltransferase are shown in Fig. 4. Two different enzymes have been identified, one acting on adenine, another on hypoxanthine and guanine. Formation of nucleotides may also occur by conversion of free bases to ribonucleosides (or deoxyribonucleosides) and subsequent phosphorylation to ribonucleotides (or deoxyribonucleotides) as shown in Fig. 5. The first step is catalyzed by nucleoside phosphorylase with base and ribose (or deoxyribose) -1-phosphate as substrates. The second step is catalyzed by nucleoside kinase in the presence of a ribonucleoside or deoxyribonucleoside triphosphate. Several nucleoside kinases have been isolated which have variable degrees of specificity for the nucleoside substrates. Although, *in vitro* conversion of purine bases to the corresponding nucleotides *via* nucleoside phosphorylase and nucleoside kinase reactions have been demonstrated, their importance *in vivo* has been questioned due to the fact that the purine nucleosides are rapidly degraded.

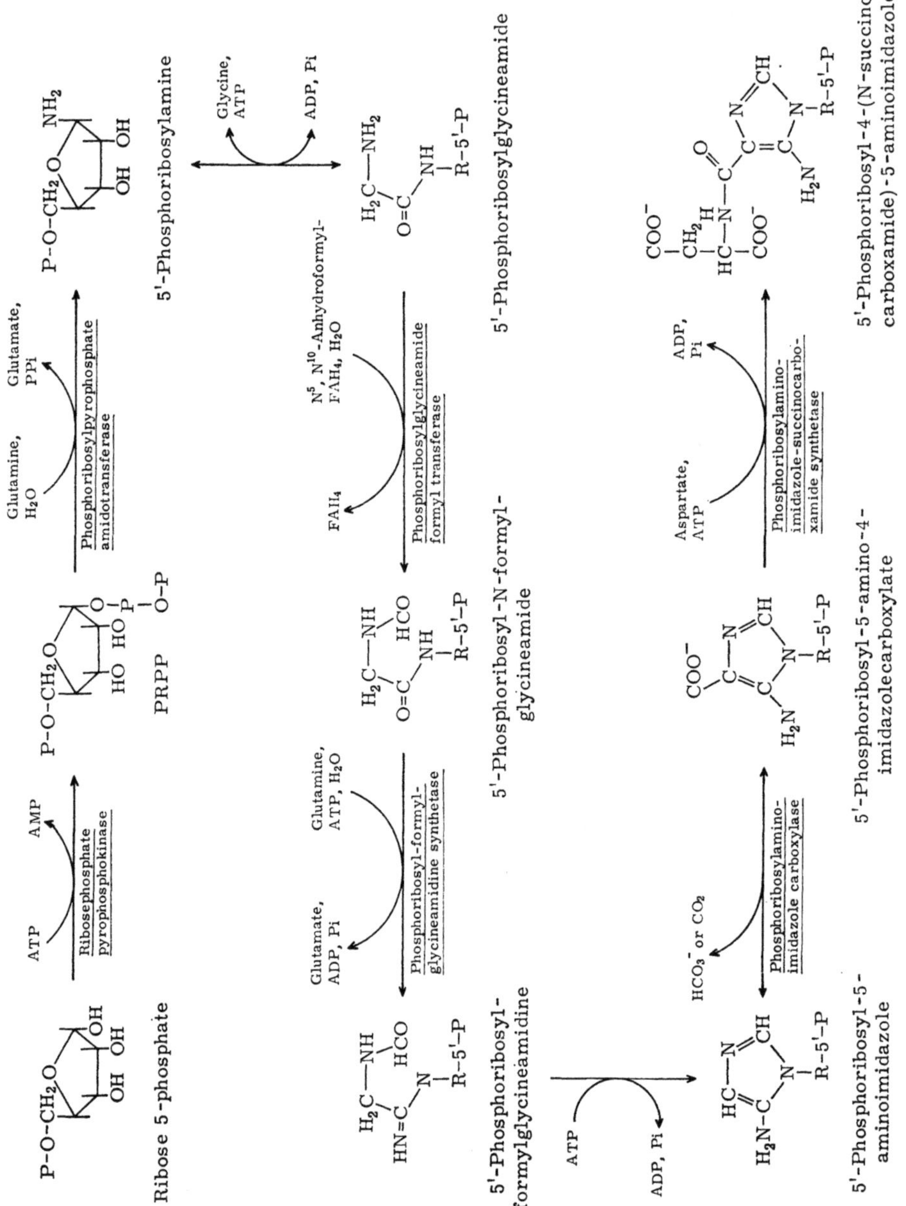

The macromolecules, DNA and RNA, are synthesized by sequential formation of phosphodiester bonds between nucleotides. The two purine deoxyribonucleoside triphosphates, dATP and dGTP, in the presence of the two pyrimidine deoxyribonucleoside triphosphates dCTP and dTTP, and DNA, are utilized for DNA synthesis in reactions catalyzed by DNA polymerase. This process of *replication* consists of an exact copying of DNA nucleotide sequence ensuring that all the cells in a multi-

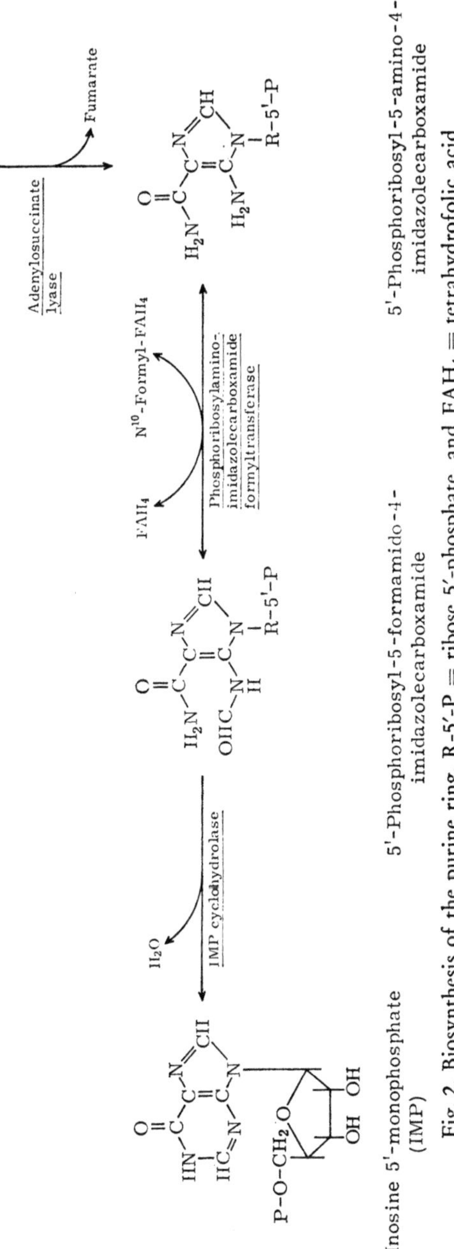

Fig. 2. Biosynthesis of the purine ring. R-5'-P = ribose 5'-phosphate, and FAH₄ = tetrahydrofolic acid

cellular organism have the same genetic information. In the biological process called *transcription*, all the four purine and pyrimidine ribonucleoside triphosphates (ATP, GTP, CTP, UTP) are utilized for RNA synthesis by the enzyme, DNA-dependent RNA polymerase. This process also involves a copying mechanism in the sense that the enzyme produces the nucleotide sequence of RNA directed by the specific sequence present in DNA. It should be noted that the mechanism of copying of a polynucleotide chain is dependent on the characteristic hydrogen-bonding properties of the bases by which adenine pairs with thymine (or uracil) and guanine with cytosine. Another biological process which is pertinent to the present discussion is termed *translation*. By this process the information in a species of RNA called messenger RNA (mRNA) is translated into the amino acid sequence of polypeptide chains. A sequence of three ribonucleotides (codon) in mRNA is read for an amino acid through the medium of a specific adaptor molecule (transfer RNA) carrying a specific amino acid and containing in a definite region of its structure a sequence of three nucleotides (anticodon) capable of hydrogen-bonding with the codon. A discussion of these three important processes *(replication, transcription* and *translation)* of macromolecular biosynthesis is beyond the scope of the present discussion. However, an understanding of the metabolism of macromolecules is required to follow the basis of metabolic action of the base analogues. A number of reviews and articles dealing with the subject of biosynthesis of DNA, RNA or protein have appeared recently. References may be made to KORNBERG (1961), MITRA and KORNBERG (1966), SUEOKA (1967), SMELLIE (1963), HURWITZ and AUGUST (1963), BAUTZ (1967), ARNSTEIN (1965), LENGYEL (1966), and to an excellent monograph by INGRAM (1965) on the biosynthesis of macromolecules.

There are enzymes in biological systems which are capable of degrading poly-nucleotides into small molecules. These small molecules may be eliminated or

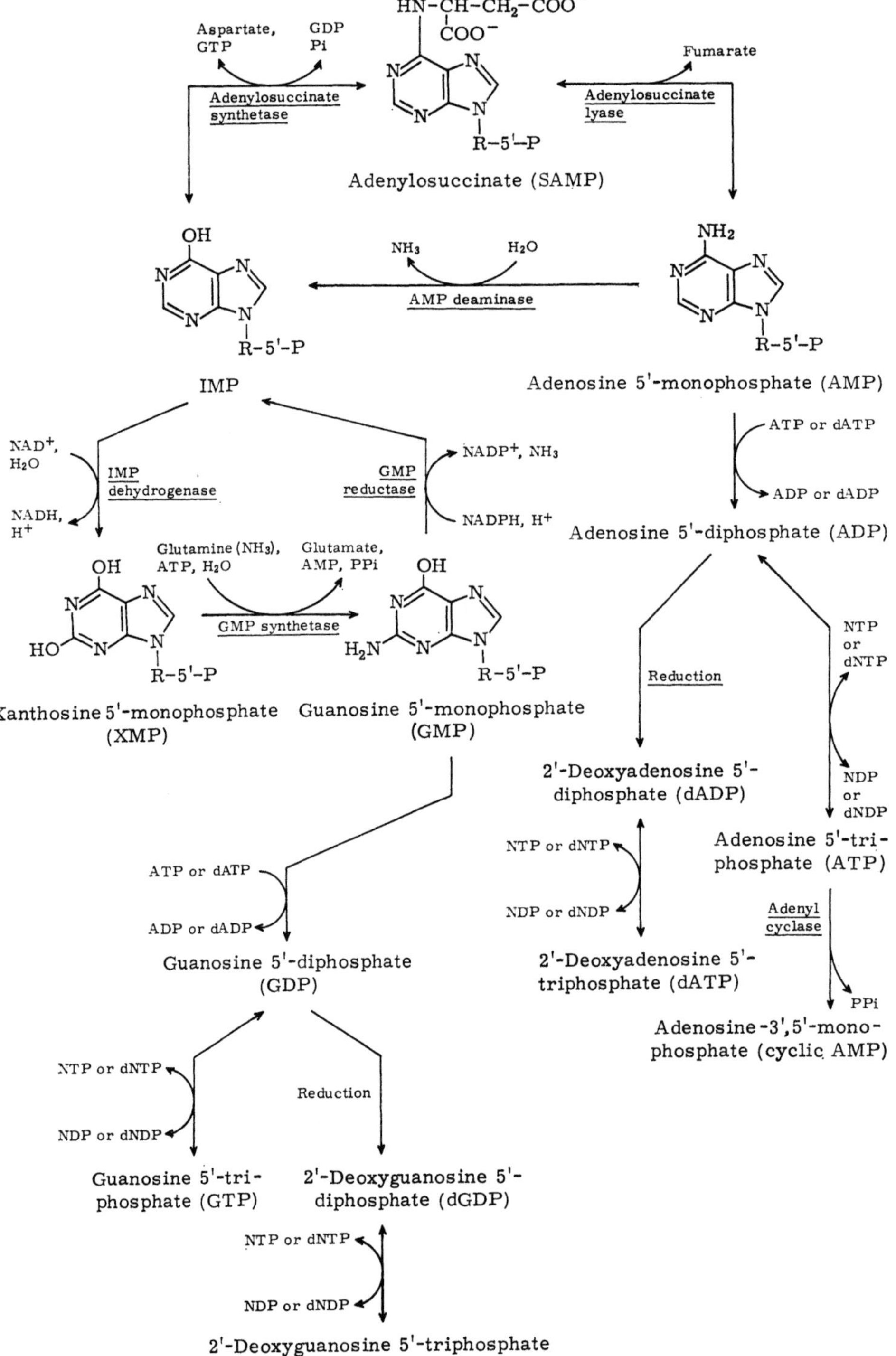

Fig. 3. Interconversions and phosphorylations of purine nucleotides.
R-5'-P = ribose 5'-phosphate; N = purine or pyrimidine nucleoside; and d = 2'-deoxy

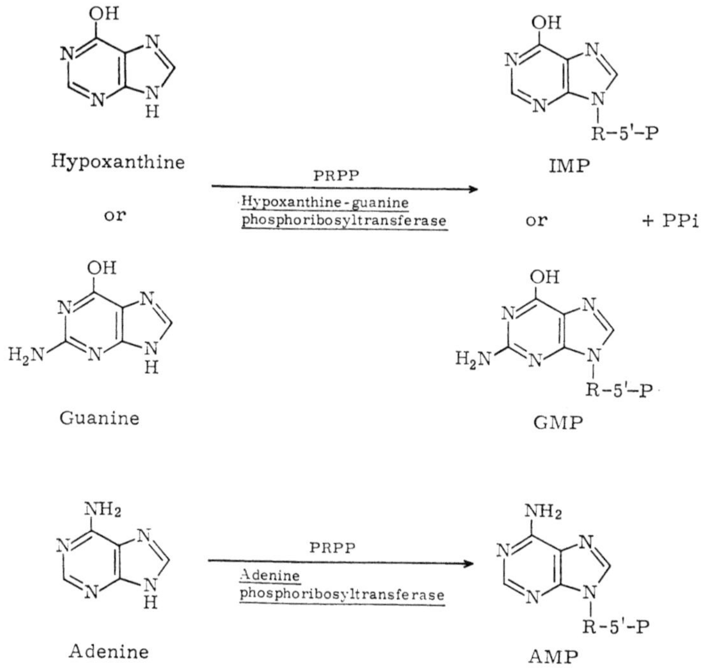

Fig. 4. Biosynthesis of purine ribonucleotides directly from purine bases

Purine base + R-1-P or dR-1-P ⇌ (Nucleoside phosphorylase) Purine ribonucleoside or Purine deoxyribonucleoside + Pi

Purine ribonucleoside or Purine deoxyribonucleoside + [NTP or dNTP] → (Nucleoside kinase) Purine ribonucleoside monophosphate or Purine deoxyribonucleoside monophosphate + [NDP or dNDP]

Fig. 5. Biosynthesis of purine nucleotides *via* purine nucleosides

reutilized in synthetic reactions. DNA and RNA are hydrolyzed to respective deoxyribo- and ribonucleotides by the action of nucleases and phosphodiesterases. The mononucleotides are split by various specific and nonspecific phosphatases to yield deoxyribo- and ribonucleosides. The nucleosides may be acted upon by nucleoside phosphorylases to yield free purine or pyrimidine bases. The free purine bases may catabolize to end products which vary widely with different animal species. The stages in the breakdown of purines are shown in Fig. 6. The end product is uric acid in man and other primates, birds, and in certain reptiles. A further oxidized product, allantoin, is excreted by mammals ohter than primates, and gastropods. Urea and glyoxylic acid are the end products in most fishes, amphibia, and fresh water

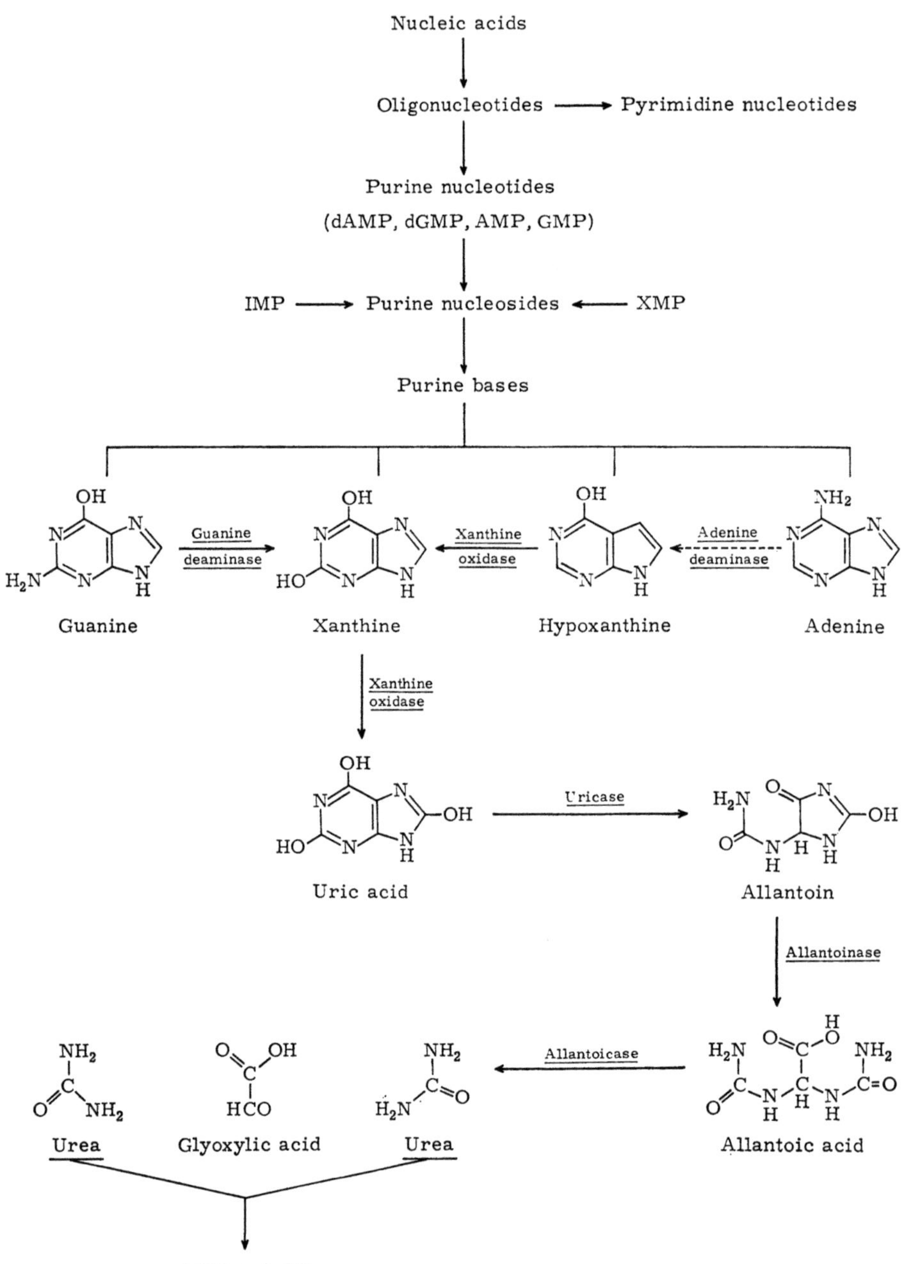

Fig. 6. Degradation of purines

lamellibranches. Finally, marine lamellibranches, crustacea, and gephyrean worms convert the purines through urea to ammonia and carbon dioxide.

In human tissues free adenine is not deaminated to hypoxanthine apparently due to the absence of the enzyme, adenine deaminase. Thus, if adenine is not anabolized to nucleoside or nucleotide, it is excreted unchanged. The enzyme, xanthine oxidase, occurs abundantly in liver and intestinal mucosa. It has been suggested that uric acid synthesis is largely a hepatic process in man because of the absence of xanthine oxidase in most tissues and its high activity in liver (WYNGAARDEN, 1966).

In addition to the biosynthetic and degradative reactions of purines, there is another consideration which is relevant to the present subject of discussion. This involves the regulation of nucleotide biosynthesis by feedback control. A summary of the reactions in which purine nucleotides participate in the regulation of their formation is outlined in Fig. 7. These controls result from inhibition or activation of

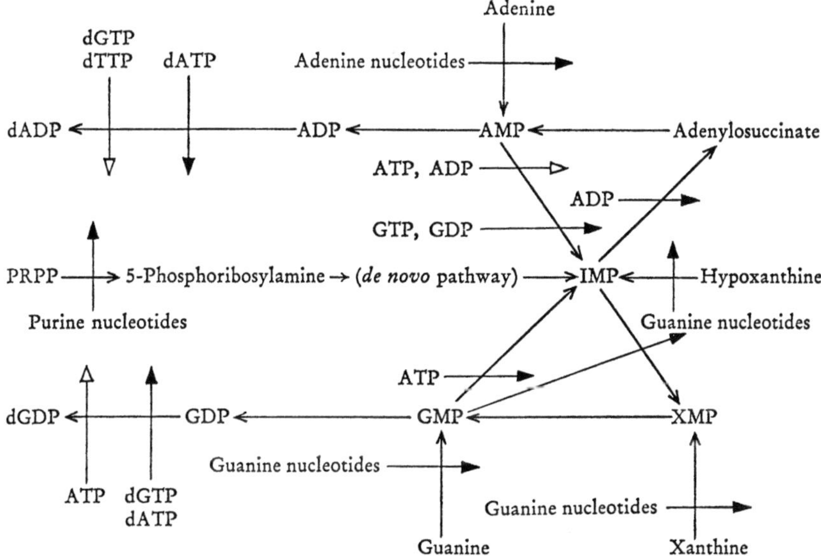

Fig. 7. Feedback controls in purine metabolism. The lines with ▶ indicate inhibition and with ▷ stimulation

certain enzymes by specific nucleotides which modify the catalytic activity of the enzymes by combining with them at sites other than the catalytic sites. Allosteric regulation of enzyme activity may sometimes be explicable in terms of modified interaction between polypeptide subunits of an enzyme or other conformational changes in the enzyme molecule resulting in either an increase or decrease in its catalytic activity. The inhibition of the early step in the *de novo* pathway of purine biosynthesis, ie., the formation of 5-phosphoribosylamine from PRPP and glutamine by purine nucleotides illustrates an important biological phenomenon by which cellular economy and metabolite levels are maintained through inhibition of the first enzyme in a sequence of reactions by the end products.

B. Mechanisms of Action of Purine Analogues

6-Mercaptopurine

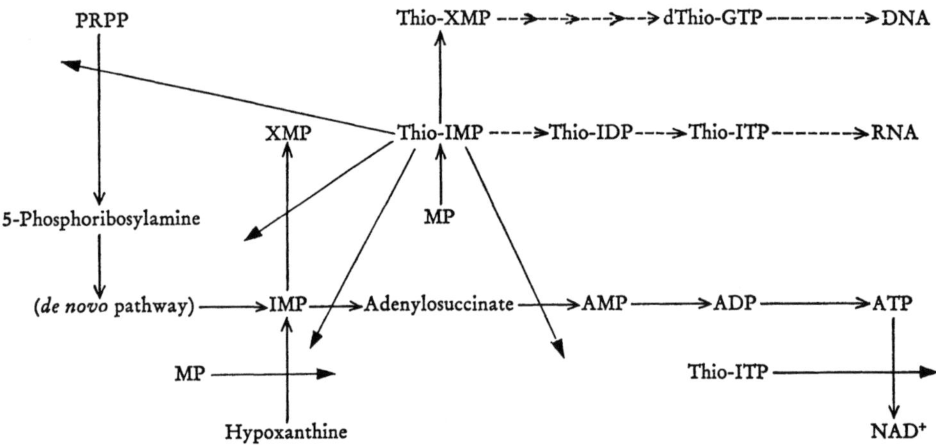

6-Mercaptopurine (MP) was synthesized in 1952 by ELION, BURGI and HITCHINGS. This purine analogue has been useful in the treatment of childhood leukemia and also with a limited success in the treatment of adult acute leukemia. The compound may be considered to be an analogue of hypoxanthine because of its close structural resemblance to that metabolite. The biochemical effects of MP have been studied extensively in bacteria, in mouse neoplasms and other tissues and correlated with inhibition of cell growth. Conversion of MP to nucleotide derivatives and the present knowledge on their inhibitory sites are outlined in Fig. 8.

Fig. 8. Metabolism of MP and the known sites of inhibition. The lines with ▶ indicate affected sites and the broken lines represent reactions which have not been definitely established

MP is converted to the ribonucleotide, thioinosinic acid (thio-IMP), by the action of guanine-hypoxanthine phosphoribosyltransferase (BROCKMAN and ANDERSON, 1963). This reaction (1)

$$MP + PRPP \xrightarrow[\text{phosphoribosyltransferase}]{\text{Guanine-hypoxanthine}} \text{thio-IMP} + PPi \tag{1}$$

is a requirement for MP to exert its inhibitory effects. WAY, DAHL and PARKS (1959) obtained evidence suggesting that thio-IMP is further phosphorylated to thio-IDP and thio-ITP by a partially purified pork kidney enzyme preparation. However, these workers did not isolate these polyphosphate derivatives of MP. Incorporation of sulfur from ^{35}S-MP into both RNA and DNA of several mouse tumors has been reported (ELION, BIEBER and HITCHINGS, 1954; BIEBER et al., 1961), although the results did not prove conclusively the occurrence of MP in the internucleotide linkage of polynucleotides. However, SCANNELL and HITCHINGS (1966) recently isolated DNA from MP-resistant adenocarcinoma 775 after treatment of the mice with 8-^{14}C-labeled MP. After enzymatic degradation of the DNA, the isotope was found to be associated mainly with deoxythioguanosine. Deoxyadenosine and deoxyguanosine contained about 20 and 10 per cent respectively, of the total radioactivity, but no evidence for the presence of deoxythioinosine was obtained. The occurrence of deoxythioguanosine in place of expected deoxythioinosine in this DNA, may be explained by the conversion of thio-IMP to thioguanylic acid (thio-GMP) prior to incorporation into DNA. It is known that thio-IMP can be oxidized to thioxanthylic acid (thio-XMP) by IMP dehydrogenase (ATKINSON, ECKERMANN and STEPHENSON, 1965; HAMPTON, 1963). Demonstration of the substrate activity of thio-XMP for the GMP synthetase to give thio-GMP, and of the conversion of thio-GMP to dthio-GTP by kinases and reductases, will be required for clear understanding of the reaction sequences.

MP is a substrate and a competitive inhibitor of guanine-hypoxanthine phosphoribosyltransferase (WAY and PARKS, 1958; ATKINSON and MURRAY, 1965). The reaction product, thio-IMP, is the most active metabolic form of MP in producing its inhibitory effects at several sites of IMP metabolism. It is a competitive inhibitor of IMP dehydrogenase isolated from Ehrlich ascites tumor cells (ATKINSON, MORTON and MURRAY, 1963). This enzyme catalyzes the conversion of IMP to XMP. HAMPTON (1963) suggested that thio-IMP binds to the reaction site of the enzyme in place of IMP and there reacts with a sulfhydryl group of the enzyme to form a stable disulfide bond. This suggestion is based on the observation that thio-IMP and IMP compete as substrates for the dehydrogenase when glutathione is present, but in the absence of glutathione thio-IMP becomes markedly more inhibitory, and the inhibition produced is reversed by glutathione but not by IMP. The demonstration of the nonenzymatic reaction of thio-IMP with mercaptans also supports the suggested mechanism. Conversion of IMP to AMP *via* adenylosuccinate (SAMP) is inhibited by thio-IMP. ATKINSON, MORTON and MURRAY (1964) studied the kinetics of thio-IMP inhibition of these two enzymes, adenylosuccinate synthetase and adenylosuccinate lyase from Ehrlich ascites tumor cells. It is a non-competitive inhibitor of the conversion of IMP to SAMP, and competitive inhibitor of the conversion of SAMP to AMP. The inhibitory effect of thio-IMP on adenylosuccinate lyase isolated from yeast is enhanced by copper ions, and the inhibition by copper is reversed by thiols other than thio-IMP, including thioinosine (BRIDGER and COHEN, 1963). This indicates that a copper-thio-IMP complex is a specific inhibitor of the enzyme. It has been suggested that the complex combines with the enzyme such that the metal ion bridges the sulfhydryl group of the enzyme and the sulfhydryl group of thio-IMP. Since the inhibition shows competitive kinetics, it is likely that the SH-group of the enzyme involved in such interaction, is located close to the site at which the amino group of SAMP is held in the enzyme substrate complex.

The formate activating enzyme, formyltetrahydrofolate synthetase which catalyzes the conversion of tetrahydrofolate to 10-formyltetrahydrofolate (reaction 2),

$$\text{Tetrahydrofolate} + \text{Formate} + \text{ATP} \rightleftharpoons \text{10-Formyltetrahydrofolate} + \text{ADP} + \text{Pi} \qquad (2)$$

is inhibited *in vitro* by MP (WILMANNS, 1962). UNGER and SILBER (1964) suggested
that MP is inhibitory to the formate activating enzyme by virtue of its resemblance
to the adenine moiety of ATP which is probably bound to an active site on the
enzyme. The activity of the enzyme present in homogenates of rat liver or samples
from chronic and acute myelocytic leukemia patients, is more strongly inhibited by
MP than by adenine, and the inhibition produced by MP is counteracted by cysteine.
The results indicate a possible disulfide bond formation between the analogue and the
enzyme.

Inactivation of GMP reductase (converting GMP to IMP), purified from *Aero-bacter aerogenes*, by thio-IMP has been reported recently (BROX and HAMPTON,
1968). Evidence has been obtained indicating that this inactivation is also related to
the formation of disulfide bond with an $-SH$ group at the GMP site.

CARBON (1962) reported that thio-IDP is a potent inhibitor of polynucleotide
phosphorylase obtained from *Micrococcus lysodeikticus*. Inhibition of polyadenylic
acid synthesis (reaction 3) by this nucleoside diphosphate analogue is dependent

$$nADP \rightleftharpoons (AMP)_n + nPi \tag{3}$$

upon the concentration of the substrate (ADP), indicating a competition between
ADP and thio-IDP for the enzyme. Thio-IDP, however, is not a substrate for poly-nucleotide phosporylase, even at high concentrations of the enzyme.

Adenyl transfer to nicotinamide mononucleotide (NMN) (reaction 4) catalyzed

$$NMN + ATP \xrightarrow{\text{NMN adenyltransferase}} NAD + PPi \tag{4}$$

by NMN adenyltransferase of liver nuclei is competitively inhibited by thio-ITP
(ATKINSON, JACKSON and MORTON, 1961). Inhibition of this reaction was thought
to have special importance in neoplasia. BRANSTER and MORTON (1956) found that
the activity per nucleus of this enzyme in a number of tumors is considerably less
than that of comparable normal tissues. The concentration of nicotinamide nucleotide
coenzymes per nucleus is also lower in tumors than in normal tissues (MORTON, 1958;
1961). On the basis of these observations MORTON (1958, 1961) proposed that in-hibition of this enzyme of the cell nucleus would preferentially block the formation
of the coenzymes in the tumor cells. Although a mention has been made of symptoms
similar to those of nicotinamide coenzymes deficiency in patients after prolonged
treatment with MP, such studies have not received much attention.

It was observed by BENNETT et al. (1963) and BROCKMAN and CHUMLEY (1965)
that an early step of the *de novo* synthesis of purines is also inhibited by the presence
of MP. These findings can be correlated with the results of McCOLLISTER et al.
(1964) showing that thio-IMP inhibits the enzymatic conversion of PRPP to 5-phos-phoribosylamine, thus mimicking the effect of purine ribonucleotides as negative
feedback inhibitors.

It is apparent that MP may exert its effect at multiple loci of nucleic acid meta-bolism. It is important to know which of these reactions are predominantly affected
in vivo during the process of growth inhibition by MP and its metabolites. Identifica-tion of such reactions depends on the biological importance of the individual enzymic
reaction, concentration of that enzyme in the cell, relative pool sizes of metabolite
and inhibitor, and on the binding affinities of the substrate and inhibitor for the
enzyme. The available data on the last criterion are shown in Fig. 9. The concentra-

tion of MP under therapeutic regimens is usually less than 10^{-5} M. Thus the enzymes for which MP or its nucleotide derivatives have a Ki value of 10^{-3} M to 10^{-4} M may not be considered to be affected *in vivo* to any significant extent (ELION, 1967). It is known that in man about sixty percent of the amount of intravenously injected MP is excreted in one to five hours. Thus it is unlikely that a sufficiently high level of MP could accumulate in tissues or blood to produce an inhibition of formate activating enzyme *(reaction VIII)*. This assumption would hold true if the *in vitro* kinetics approximate those *in vivo* unless a remarkable concentrating selectivity for

Number	*Reaction*	*Inhibitor*	*Km*	*Ki*
I.	IMP → SAMP	Thio-IMP	3.0×10^{-5} M	3.0×10^{-4} M
II.	IMP → XMP	Thio-IMP	1.4×10^{-5} M	3.6×10^{-6} M
III.	Hypoxanthine → IMP	MP	1.1×10^{-5} M	8.3×10^{-6} M
IV.	nADP → (AMP)$_n$	Thio-IDP	1.7×10^{-3} M*	3.3×10^{-5} M*
V.	PRPP → Phosphoribosylamine	Thio-IMP	—	4.4×10^{-5} M
VI.	ATP+NMN → NAD	Thio-ITP	7.4×10^{-5} M	5.0×10^{-5} M
VII.	SAMP → AMP	Thio-IMP	2.8×10^{-6} M	3.0×10^{-4} M
VIII.	Tetrahydrofolate → Formyltetrahydrofolate	MP	5.0×10^{-4} M	3.3×10^{-3} M 2.5×10^{-4} M

Fig. 9. Binding constants of MP and its nucleotide derivatives in the inhibited reactions (* the values are actually concentrations for 50% inhibition) (ELION, 1967)

the analogue is present at the target site (UNGER and SILBER, 1964). Although the binding constants for the substrate and the inhibitor are almost identical in *reaction VI*, the metabolic pool of ATP is much greater than that of thio-ITP. It appears, therefore, that this reaction would have minor contribution to the overall effect. Requirement of IMP for growth is maintained mainly by its *de novo* synthesis. As such, inhibition of IMP formation from hypoxanthine *(reaction III)* is not likely to interfere with cell survival to any significant extent unless this reaction has any other important part in the control of cellular processes. *Reaction IV* is catalyzed by polynucleotide phosphorylase which is believed to be a degradative enzyme *in vivo*. Inhibition of this enzyme would not contribute much towards the total effect of MP and nucleotides.

From the reactions which have been studied so far it can be pointed out that the reactions catalyzed by adenylosuccinate synthetase, IMP dehydrogenase and phosphoribosylpyrophosphate amidotransferanse *(reactions I, II, and V)* are most likely to be affected subsequent to MP administration. For *reaction I*, the binding constant of thio-IMP (Ki) is ten times that of IMP (Km). But as IMP is present in very low concentrations in cells, it has been suggested that the competition may be adequate to explain the inhibition of cell division (ELION, 1967). As the Ki is lower than Km in *reaction II*, the reaction will be highly susceptible to inhibition. Interference with the conversion of PRPP to 5-phosphoribosylamine *(reaction V)* by thio-IMP through a feedback mechanism has been observed in experimental neoplasms (BENNETT et al., 1963; BROCKMAN and CHUMLEY, 1965; McCOLLISTER et al., 1964). Since the Ki value for this reaction is low, this reaction may also be a major site of attack. However, some doubt about the importance of this effect remains since HITCHINGS and ELION (1967) reported that the extent of feedback inhibition by MP (as thio-IMP) in a MP-resistant line of adenocarcinoma 755 is the same as in the sensitive line.

2*

Resistance: Resistance to MP in experimental neoplasms is frequently accompanied by loss of guanine-hypoxanthine phosphoribosyltransferase, the enzyme responsible for the conversion of MP to its active metabolite, thio-IMP. However, there may be several other different ways for the development of resistance against MP. For example, cells of a subline of Ehrlich ascites carcinoma, which are resistant to the analogue, do not convert MP to thio-IMP. But guanine-hypoxanthine phosphoribosyl-transferase is as active in extracts of these resistant cells as in extracts of the MP-sensitive parent line of tumor cells (WHITE, 1959). These results indicate the possibility of formation of the transferase with an altered affinity. SCOTT, MARINO and GABOR (1966) reported that the leukocytes of two cases of human acute myeloblastic leukemia with acquired drug-resistance showed an increased capacity for *in vitro* purine synthesis and suggested that an increase in purine biosynthetic enzyme reserves may be a mechanism of MP resistance in some human leukemias. The other possible mechanisms which have been considered are decreased permeability to MP, or detoxication of the inhibitor through cleavage of sulfur (BROCKMAN and ANDERSON, 1963).

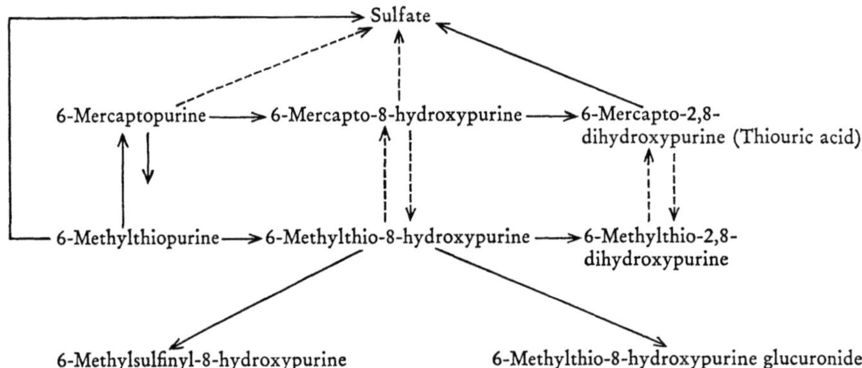

Fig. 10. Catabolic pathways of MP. Broken lines represent possible routes of conversion not definitely established (ELION, 1967)

Catabolism: The catabolic pathways of MP are summarized in Fig. 10. Similar to the oxidation of hypoxanthine to uric acid, the analogue is oxidized by xanthine oxidase to thiouric acid. The probable intermediate of this conversion is 6-mercapto-8-hydroxypurine. It is also known to be 5-methylated with subsequent oxidative reactions of the methylated derivatives to various products as shown in Fig. 10. The metabolic fate of MP is similar in mouse and man with respect to the amounts of free MP and thiouric acid excreted. However, as regards the excretion of 6-methylsulfinyl-8-hydroxypurine, it is the predominent metabolite of the methylthiopurines, and in mouse this pathway of inactivation is minor (ELION et al., 1963). Thiouric acid gives rise to sulfate in the mouse after cleavage by uricase. The intermediate steps involved in sulfate formation have not been investigated. Since man lacks uricase, sulfate cannot arise *via* thiouric acid. The finding that 6-methylthiopurine is converted more extensively to sulfate than is MP, together with the identification of 5-methylated products from MP, suggests that in man a major route from MP to sulfate is *via* 5-methylation (ELION, RUNDLES and HITCHINGS, 1964). This suggestion is consistent with the observations that the formation of sulfate is not affected by xanthine oxidase inhibition in man, but is inhibited in mouse through prevention of thiouric acid formation (ELION, RUNDLES and HITCHINGS, 1964).

It is apparent from these catabolic reactions of MP that suppression of oxidative degradation of the inhibitor would result in potentiation of its inhibitory effects. The xanthine oxidase inhibitor, 4-hydroxy (3,4-d) pyrazolopyrimidine (allopurinol) has shown the expected effect. Administration of this compound concurrently with MP results in a marked decrease in the oxidation of the latter to thiouric acid (ELION et al., 1963 a), and thus induces a several fold potentiation of MP as inhibitor of neoplastic growths or immune responses (ELION and HITCHINGS, 1965). Allopurinol, when tested for its activity alone has little effect on experimental tumors (WHITE, 1959; SHAW et al., 1960).

6-[(1-Methyl-4-nitro-5-imidazolyl)thio]purine (Imuran or Azathioprine)

Imuran

6-[(1-Methyl-4-nitro-5-imidazolyl)thio]purine, also known as Imuran or Aza-thioprine, was synthesized with a view to protect the sulfhydryl group of 6-mercapto-purine (MP) from rapid methylation followed by oxidation of the methylated derivative. This derivative of MP has been found to be active against a variety of transplantable tumors (ELION et al., 1961) and has shown therapeutic benefit in some patients with autoimmune and hypersensitivity diseases (RUNDLES et al., 1961). Activity of Imuran is the same whether the drug is given intraperitoneally or orally, but the chemotherapeutic index is superior with oral administration. Metabolic studies have shown that Imuran is well absorbed from the intestinal tract and that it is converted *in vivo* in mouse, dog or man to free MP (ELION et al., 1961). The breakdown (reaction 5) of this compound *in vivo* and *in vitro* has been demonstrated to occur by reacting with sulfhydryl compounds, such as glutathione, cysteine, or

Imuran   +   SH⁻   ⟶   MP   +   (5)

some proteins (ELION et al., 1961; BRESNICK, 1959). Cleavage of Imuran by liver homogenates is not reduced by heating the homogenates to 100° for 15 minutes, and MP formation occurs with trichloroacetic acid-precipitable liver protein as well as with the supernatant fraction thereform (BRESNICK, 1959). These results show that the breakdown process is nonenzymatic in nature. Mild performic acid treatment of the reaction mixture blocks the formation of MP indicating the involvement of $-SH$ group. The metabolic fate of the released imidazole moiety is not known.

The mechanisms of action of Imuran in growth inhibition are those as outlined in Fig. 8 under MP. The toxic symptoms produced by Imuran are qualitatively similar to those produced by MP, but apparently there is no damage to the intestinal mucosa at doses which depress the bone marrow (ELION et al., 1961). The immunosuppressive effects of Imuran and MP are different in experimental systems. The dose of MP required to suppress the antibody formation is 0.25 mmole/kg whereas Imuran is active at a much lower dose, approximating 0.07 mmole/kg. When the mice were pretreated with glutathione prior to the administration of Imuran, so that splitting to MP occurs much more rapidly, then the dose response curve is identical to that with MP (ELION, 1967). On the basis of these data, ELION (1967) suggests that the slow release of MP and simultaneous binding of sulfhydryl groups may each contribute to the superiority shown by Imuran in immunosuppression. Similar advantage of Imuran was also observed in dog kidney homografts (CALNE, 1961; CALNE, ALEXANDRE and MURRAY, 1962). However, the therapeutic effectiveness of this analogue over MP has not yet been established in human subjects.

9-β-D-Ribofuranosyl-6-methylthiopurine (6-Methylthiopurine ribonucleoside)

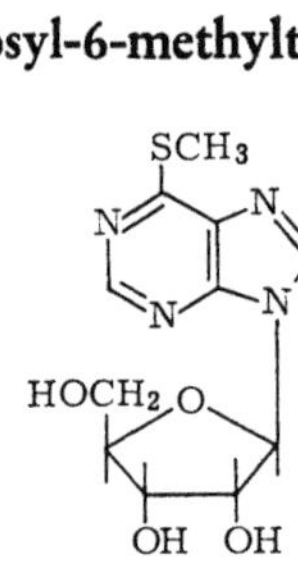

6-Methylthiopurine ribonucleoside 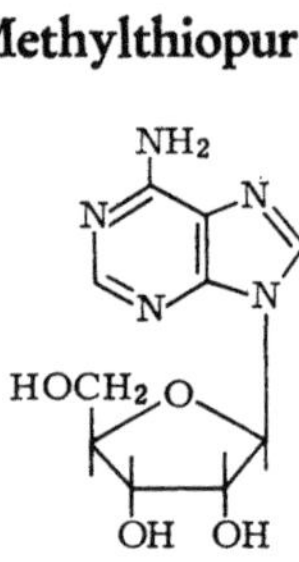   Adenosine

9-β-D-Ribofuranosyl-6-methylthiopurine (6-methylthiopurine ribonucleoside or MeMPR), a derivative of 6-mercaptopurine ribonucleoside, is a potent inhibitor of the growth of several mouse tumors (PATERSON, 1961) and is effective against some tumors which have become resistant to 6-mercaptopurine (MP) therapy (BENNETT et al., 1965). A combination of MP and MeMPR showed significant therapeutic effects on experimental neoplasms *in vivo* (SCHABEL, LASTER and SKIPPER, 1967; WANG, SIMPSON and PATERSON, 1967; PATERSON and WANG, 1968). PATERSON and WANG (1968) reported that the combination therapy is more inhibitory to the Ehrlich ascites carcinoma than the sum of the independent drug effects. In whole cells or in isolated enzyme systems MeMPR does not act as a substrate for nucleoside phosphorylases. But the analogue is a substrate for a nucleoside kinase and is extensively converted to 6-methylthiopurine ribonucleoside 5'-phosphate which may also be named 6-methyl-thioinosine 5'-monophosphate (methylthio-IMP) (BENNETT et al., 1965; CALDWELL, HENDERSON and PATERSON, 1966; SCHNEBLI, HILL and BENNETT, 1967). Its conversion to the nucleotide in cells which fail to convert 6-mercaptopurine ribonucleoside to the nucleotide form shows MeMPR to be a substrate for a nucleoside kinase which does not act on the ribonucleoside of MP (BENNETT et al., 1965). CALDWELL, HENDERSON and PATERSON (1966) obtained evidence indicating that the phosphorylation of MeMPR in tumor cells is mediated *(in vitro* or *in vivo)* by adenosine kinase and requires ATP as phosphate donor. SCHNEBLI, HILL and BENNETT (1967) purified adenosine kinase from human tumor cells in culture and found that the enzyme phosphorylates adenosine or MeMPR in the presence of a ribonucleoside triphosphate. The ribosyl derivative of MP is not a substrate for the enzyme, but it inhibits the phosphorylation of both adenosine and MeMPR. The differences in the substrate activity of these two ribonucleosides for adenosine kinase is of considerable interest. It has been pointed out (Fig. 11) that, whereas inosine (I) and MP ribo-

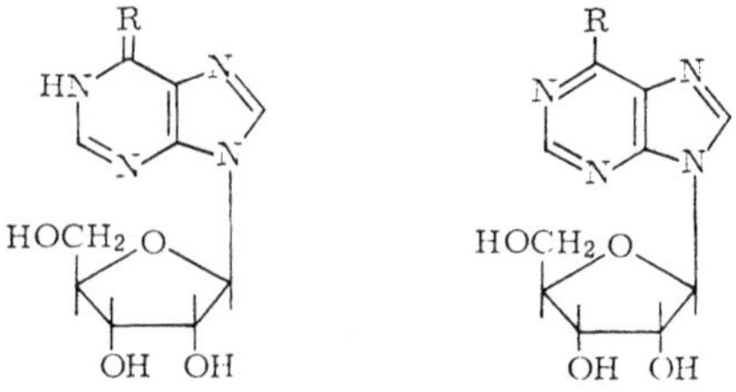

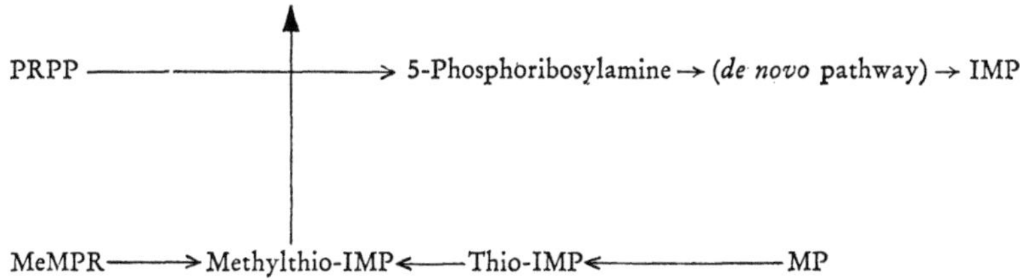

Fig. 11. Predominant tautomeric forms of inosine, adenosine and mercaptopurine ribonucleosides

nucleoside (II) both bear a proton at N-1 (since they exist largely in the lactam and thiolactam forms), adenosine (III) and MeMPR (IV) do not (BENNETT et al., 1965). The absence of a proton at N-1 of the purine ring might, therefore, be more critical for activity as a substrate for the kinase than the nature of the group at the 6-position of the ring (BENNETT et al., 1965). On this basis inosine kinase may phosphorylate MP ribonucleoside. However, it should be noted that inosine or guanosine kinases have not been reported to occur in mammalian cells.

All facets of the metabolism and inhibitory action of MeMPR are not known; however, it can be said at present that methylthio-IMP is the active form of the inhibitor (Fig. 12) and that no other major metabolites of MeMPR are formed at least

PRPP ——————————————→ 5-Phosphoribosylamine → *(de novo* pathway) → IMP

MeMPR ———→ Methylthio-IMP ←——— Thio-IMP ←——————— MP

Fig. 12. Phosphorylation of MeMPR to the active metabolite and the known site of inhibition as indicated by ▶

in mouse tissues, Ehrlich ascites carcinoma cells or in cell-free extracts of tumors (CALDWELL, HENDERSON and PATERSON, 1966). The active inhibitor, methylthio-IMP is also formed in small amounts *in vivo* by methylation of thio-IMP, a product of MP administration (ALLAN, SCHNEBLI and BENNETT, 1966). The primary site of action

of methylthio-IMP is in the feedback inhibition of *de novo* purine synthesis (HENDER-
SON, 1963; BENNETT et al., 1965). Recent work has shown that methylthio-IMP is a
more potent inhibitor of phosphoribosylpyrophosphate amidotransferase than is thio-
IMP (BROCKMAN, 1968) (Fig. 12). PATERSON and WANG (1968) have carried out an
interesting study on the influence of MeMPR upon the metabolism of MP and *vice
versa* in Ehrlich ascites carcinoma cells *in vivo*. They observed that pretreatment with
MeMPR greatly increases the formation of nucleotides of MP; however, similar treatment with MP does not change the rate of phosphorylation of MeMPR by the tumor
cells. This MeMPR-induced stimulation of MP anabolism is apparently related to
inhibition of phosphoribosylpyrophosphate amidotransferase by methylthio-IMP
accompanied by accummulation of increased concentration of PRPP. Since methylthio-IMP is an analogue of AMP, it may interfere with other reactions involving
formation or participation of AMP. Such effects as well as any possible incorporation
of the analogue into nucleic acids remain to be investigated.

Resistance: MeMPR-resistant sublines have been isolated from Ehrlich ascites
carcinoma (CALDWELL, HENDERSON and PATERSON, 1967; HO and FREI, 1968), and
from H. Ep. No. 2 cells (BENNETT et al., 1966), a line of cultured cells derived from
a human epidermoid carcinoma. In both instances, the MeMPR-selected sublines are
deficient in adenosine kinase and therefore are unable to convert MeMPR to its
nucleotide derivative. However, the mechanism of resistance formation appears to be
different in a resistant subline of Ehrlich ascites carcinoma designated EAC-RI. The
sole metabolite of MeMPR in both the parent line and EAC-RI is methylthio-IMP,
and both form the nucleotide at almost identical rates (HENDERSON, CALDWELL and
PATERSON, 1967). But the analogue is less effective as feedback inhitor of purine biosynthesis *de novo* in EAC-RI than in the sensitive cells, whereas adenine (as nucleotides)
is equally effective in either. Also the phosphoribosylpyrophosphate amidotransferase
activity of the resistant cells is inhibited by methylthio-IMP to a lesser extent than
that of sensitive parent tumor cells. From these observations, HENDERSON, CALDWELL
and PATERSON (1967) suggested that the EAC-RI cells possess an altered amidotransferase which binds methylthio-IMP less well, or for which such binding has less inhibitory effect than for the enzyme in sensitive cells.

2-Amino-6-mercaptopurine (6-Thioguanine)

6-Thioguanine

Guanine

The guanine analogue, 6-thioguanine (thio-G) possesses growth-inhibitory activity
against transplanted tumors in animals and neoplasms in man. It has therapeutic
effects similar to those of 6-mercaptopurine in animal tumors and in human leukemias
(ELION and HITCHINGS, 1965).

In order to exert its antitumor action thio-G must first be converted to its active
metabolite, 6-thioguanylic acid (thio-GMP). Enzymatic formation of thio-GMP from
thio-G and PRPP by guanine-hypoxanthine phosphoribosyltransferase has been

demonstrated (BROCKMAN, 1963). This nucleotide analogue has been reported to accumulate in much higher levels in some tumor cells than in normal tissues (MOORE and LEPAGE, 1958). It has a low capacity to act as the substrate for a nucleotide monophosphokinase, isolated from hog brain tissues or sarcoma 180 cells, which catalyzes the phosphorylation of GMP to GDP (MEICH et al., 1967). The reaction rate is less than one-hundredth of the rate obtained with GMP. However, work from laboratories of LEPAGE and colleagues (SARTORELLI, LEPAGE and MOORE, 1958; of thio-G into DNA and RNA of neoplastic cells. These results imply that thio-GDP, thio-GTP and their deoxyribonucleotide counterparts are formed in those intact cells. LEPAGE, 1960; LEPAGE and JONES, 1961) has shown small amounts of incorporation On the other hand, treatment of mice bearing implants of sarcoma 180 cells with thio-G results in the accumulation of considerable amounts of thio-GMP and only trace amounts of either di- or triphosphates of the analogue in those cells (SARTORELLI et al., 1964; BIEBER and SARTORELLI, 1964).

Thio-GMP exerts its inhibitory effects on a number of enzymes (Fig. 13). McCOL-LISTER et al. (1964) reported that the nucleotide analogue acts as a feedback inhibitor

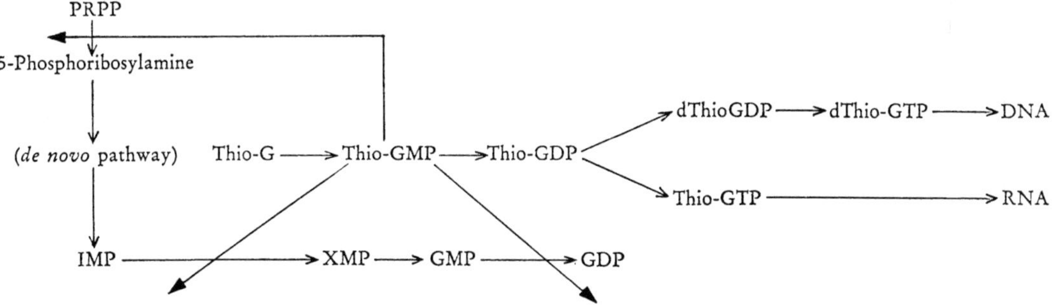

Fig. 13. Metabolism of thio-G and the known sites of inhibition. The lines with ▶ indicate affected sites

of phosphoribosylpyrophosphate amidotransferase, the first enzyme in the *de novo* pathway of purine synthesis, the *Ki* value being 2×10^{-4} M. With the accumulation of thio-GMP a significant inhibition of this enzyme *in vivo* has also been observed (SARTORELLI and LEPAGE, 1958). IMP dehydrogenase (catalyzing IMP to XMP conversion) of *Aerobacter aerogenes* (HAMPTON, 1963) or sarcoma 180 ascites cells (MEICH et al., 1967) is subject to inhibition by thio-GMP. The inhibition produced is not reversed by dialysis of the enzyme-inhibitor complex, partially reversed by glutathione, and completely reversed by treatment with dithiothreitol indicating that a covalent bond is formed between the nucleotide analogue and IMP dehydrogenase, possibly through disulfide bond formation. A nucleotide monophosphokinase, specific for the phosphorylation of GMP to GDP and isolated from hog brain, has been found to be inhibited by thio-GMP. The inhibition of this kinase is competitive and the *Ki* has a relatively low value, $5 - 8 \times 10^{-5}$ M (MEICH and PARKS, 1964; MEICH et al., 1967). Thus, one might expect that the intracellular accumulation of the thio-nucleotide would cause some decrease in the formation of all purine nucleotides by its allosteric effect on phosphoribosylpyrophosphate amidotransferase and particularly would affect the biosynthesis of GMP and its higher phosphates or its deoxynucleotides by inhibition of IMP dehydrogenase and the nucleotide monophosphokinase. Since GTP is a specific coenzyme in a number of vital metabolic reactions, and since GTP and dGTP are utilized for synthesis of nucleic acids, depletion of these nucleotides would result in a marked depression in cellular metabolism.

Interference with the above cellular reactions by thio-GMP may be responsible for growth inhibition. However, it is also possible that incorporation of the analogue into nucleic acids may have some role in the overall effect. Small amounts of incorporation of thio-GMP into both DNA and RNA of tumors have been observed and it has been suggested that the susceptibility of the tumors to the growth inhibitory activity of thio-G is directly related to the degree of incorporation of the analogue into DNA (LePage, 1960). Mecca lymphosarcoma cells are resistant to exposure to thio-G, although large amounts of thio-GMP accumulate and there is little incoporation of the analogue into DNA (LePage, Junga and Bowman, 1964). However, these neoplastic cells are susceptible to 2′-deoxythioguanosine and treatment with this derivative results in a significant incorporation of the compound into nucleic acids (LePage, Junga and Bowman, 1964). These observations may be indicative of inhibition of cellular metabolism of certain growths through incorporation of thio-G into functional macromolecules.

Resistance: Thio-G resistant L1210 leukemia cells and certain sublines of Ehrlich ascites cells resistant to the drug, exhibit significant loss of guanine-hypoxanthine phosphoribosyltransferase activity accompanied by decreased capacity to form thio-GMP (Stutts and Brockman, 1963; Ellis and LePage, 1963). Several ascites cell tumors which are insensitive to the analogue, are known to possess a kinase system for the conversion of thioguanosine or deoxythioguanosine to the nucleotide form and respond somewhat to their effects. Likewise, a resistant tumor having a poor capacity to form ribonucleotide respond to the deoxynucleoside derivative only (LePage, Junga and Bowman, 1964).

Catabolism: The catabolism of thio-G in man, as well as in lower animals, follows the general pathway shown in Fig. 14 (Elion, 1967). There are some quantitative

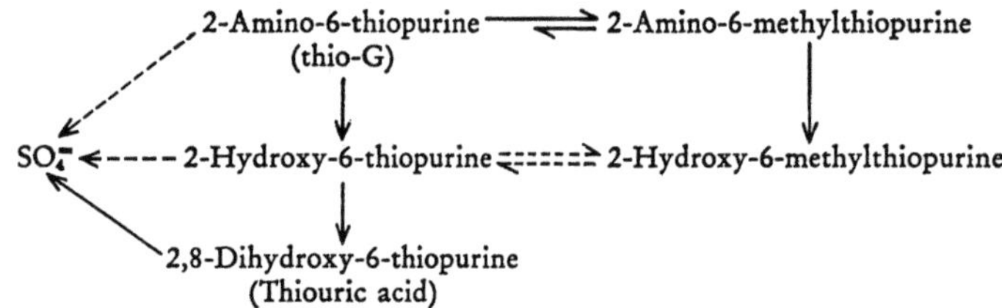

Fig. 14. Catabolic pathways of thio-G. Broken lines represent possible routes of conversion not definitely established (Elion, 1967)

differences in these catabolic reactions in man and mouse. In man there is much more methylation of thio-G than in mouse (Elion et al., 1962), while deamination and subsequent oxidation to thiouric acid occurs to only a small extent in man, but is the main pathway in the mouse (Elion, 1967). Similar to the catabolism of 6-mercaptopurine, sulfate is a major end product in the degradative reactions of thio-G. The exact route of sulfate formation is not known.

4-Hydroxy (3,4-d) pyrazolopyrimidine (Allopurinol)

Allopurinol   Hypoxanthine

4-Hydroxy (3,4-d) pyrazolopyrimidine (allopurinol or Zyloprim), a synthetic isomer of hypoxanthine was first synthesized by ROBINS (1956). The compound has been found to be a potent inhibitor of the enzyme, xanthine oxidase, which catalyzes the oxidation of hypoxanthine to xanthine and xanthine to uric acid (FEIGELSON, DAVIDSON and ROBINS, 1957; ELION et al., 1963 a). This inhibitory property of allopurinol has been further investigated for the purpose of suppressing the oxidative degradation of 6-mercaptopurine in the treatment of leukemia (ELION et al., 1962; ELION et al., 1963 a). The analogue inhibits the conversion of 6-mercaptopurine to thiouric acid in mouse and man (ELION et al., 1963 a), produces a several fold potentiation in antitumor and antiimmune effects of thiopurines in mice (ELION et al., 1962; ELION et al., 1963; ELION et al., 1963 a; ELION, MUELLER and HITCHINGS, 1959; SILBERMAN and WYNGAARDEN, 1961), and it has been used as adjunct therapy in human leukemia (RUNDLES et al., 1963). Allopurinol when tested for its activity alone showed little effect on experimental tumors (WHITE, 1959, SHAW et al., 1960). However, it was observed that the analogue is capable of decreasing serum and urinary levels of uric acid in patients with leukemia and hyperuricemia (RUNDLES et al., 1963). These observations immediately suggested a trial of the drug in gout.

Gout is considered to be an inborn error of metabolism. In acute cases of hyperuricemia, crystals of urate are formed from supersaturated body fluids, and deposits of urate may occur in joints and other tissues. Although defects in disposition of uric acid may be responsible for hyperuricemia of some patients, it has been documented that *de novo* synthesis of purines occurs at an increased rate in gouty patients associated with overproduction of uric acid (SEEGMILLER et al., 1961) and that in affected members of certain families excessive production of purines is the only cause for hyperuricemia (SEEGMILLER, 1962). Increased rate of purine synthesis may be directly related to increased catabolic reactions with accumulation of end product. Rundles et al. (1963) and WYNGAARDEN et al. (1963) were the first investigators to use allopurinol for clinical effects and they obtained indications that hyperuricemia can be controlled by this drug. Since then the analogue has been extremely useful in the treatment of gout and of hyperuricemias resulting from malignant disease, polycythemia vera or glycogen storage disease (RUNDLES, ELION and HITCHINGS, 1966; WYNGAARDEN, 1966). It has also shown usefulness in the control of uric acid stone formation in patients with hyperuricosuria (ANDERSON et al., 1967). Allopurinol is a well-tolerated drug and there has been no apparent adverse effect on renal, hepatic, central nervous or hematopoietic function (RUNDELS, METZ and SILBERMAN, 1966; WYNGAARDEN, 1966).

The sites of action and the metabolic fate of allopurinol is outlined in Fig. 15. The analogue is both an inhibitor of, and a substrate for xanthine oxidase. The major metabolite of allopurinol in mice, dogs, and human subjects is its oxidation product, oxoallopurinol [4,6-dihydroxy (3,4-d) pyrazolopyrimidine] (ELION et al., 1966).

Conversion of allopurinol to oxoallopurinol is catalyzed by xanthine oxidase, and oxoallopurinol is also an inhibitor of the enzyme (ELION, TAYLOR and HITCHINGS, 1964). Both of these forms of the inhibitor are cleared rapidly by mouse and dog kidney. However, in man the rate of renal clearance of allopurinol is much faster than

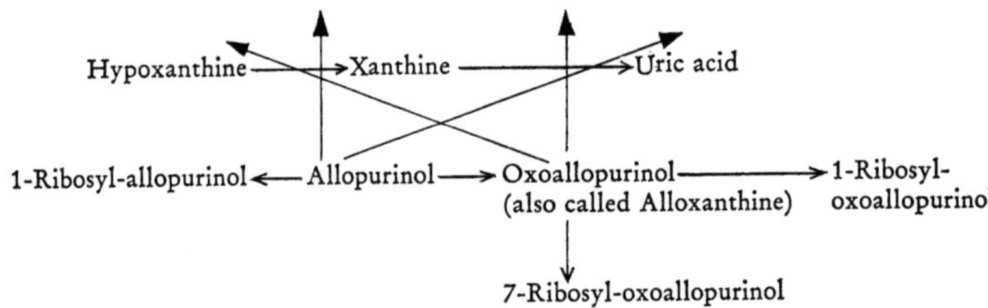

Fig. 15. Metabolism of allopurinol and its *loci* of action. (The lines with ▶ indicate sites of inhibition)

that of oxoallopurinol. It is possible that the low clearance rate of oxoallopurinol has a significant role in the therapeutic effects of allopurinol because of the clinical observations that the effectiveness of the drug increases over several days when therapy is initiated and persists for a like period when therapy is discontinued (ELION et al., 1966).

Both allopurinol and oxoallopurinol are converted to the 1-ribosyl derivatives by purine nucleoside phosphorylase (ELION, KOVENSKY and HITCHINGS, 1966; KRENITSKY et al., 1967). Allopurinol is a much better substrate than oxoallopurinol in this *in vitro* reaction, and 1-ribosylallopurinol, but not 1-ribosyloxoallopurinol, has been detected in the urinary excretions of patients treated with allopurinol (KRENITSKY et al., 1967). Oxoallopurinol has also been shown to be converted to the 7-ribosyl derivative by guinea pig intestinal preparations of uridine phosphorylase (KRENITSKY et al., 1967). Isolation of 7-ribosyloxoallopurinol from the urinary excretion of a patient treated with allopurinol seems to indicate this reaction *in vivo*. The resemblance of the configuration of oxoallopurinol to that of 2,6-dioxopyrimidine (or uracil) moiety probably determines its susceptibility to ribosylation at position 7 by uridine phosphorylase (KRENITSKY et al., 1967). Whether these nucleosides of allopurinol and oxoallopurinol can undergo phosphorylation to form nucleotides is not known. Although allopurinol is a substrate for guanine-hypoxanthine phosphoribosyltransferase (McCOLLISTER et al., 1964; WAY and PARKS, 1958), no detectable amount of nucleotide formation or incorporation into nucleic acids has been observed in experiments *in vivo* with ^{14}C-allopurinol (ELION, KOVENSKY and HITCHINGS, 1966).

8-Azaguanine

8-Azaguanine   Guanine

The base analogue azaguanine (aza-G) which contains a nitrogen atom at position 8 in place of a carbon of guanine, was synthesized in 1945 (ROBLIN et al.). The analogue is an inhibitor of multiplication of several animal and plant viruses, protozoa like *Tetrahymena gelii*, and microorganisms (ELION and HITCHINGS, 1965). It shows activity against various experimental tumors and the inhibition produced varies inversely with the guanine deaminase activity in the tumor (HIRSCHBERG, KREAM and GELLHORN, 1952). The enzyme guanine deaminase deaminates the compound to the apparently inactive 8-azaxanthine (reaction 6). Aza-G has not been useful in clinical

$$\text{Guanine} \xrightarrow{\text{Guanine deaminase}} \text{xanthine} \tag{6}$$

$$\text{Aza-G} \xrightarrow{\text{Guanine deaminase}} \text{Azaxanthine}$$

cancer chemotherapy. However, some objective responses have been reported in patients with head and neck cancer after continuous intra-arterial infusion (HALL et al., 1962).

Enzymatic conversions of aza-G to aza-GMP, aza-GDP and aza-GTP have been demonstrated (MITCHELL, SKIPPER and BENNETT, 1950; MANDEL, CARLO and SMITH, 1954; SMITH and MATHEWS, 1957; MANDEL and MARKHAM, 1958; WAY and PARKS, 1958; BROCKMAN et al., 1959; WAY, DAHL and PARKS, 1959). Correlations between sensitivity to aza-G and the capacity to form aza-GMP in tumors and *S. faecalis* have indicated that the analogue must be converted to the nucleotide derivative in order to exert its inhibitory effect (BROCKMAN, SPARKS and SIMPSON, 1957; BROCKMAN et al., 1959). The analogue is incorporated into bacterial, viral, and mammalian RNA, replacing up to 5% of the guanine residues (LASNITZKI, MATHEWS and SMITH, 1954) and aza-GTP partially substitutes for GTP as substrate for RNA polymerase (KAHAN and HURWITZ, 1962). In *B. cereus* a much higher percent (about 40%) of total RNA guanine is replaced by aza-G (SMITH and MATHEWS, 1957). Preferential incorporation of aza-G into the sRNA species has been observed in several laboratories (MANDEL and MARKHAM, 1958; LEVIN, 1963; KARON et al., 1965) and recently evidence has been obtained indicating incorporation into certain species of RNA, probably mRNA, which are essential to the integrity of the polyribosome (ZIMMERMAN and GREENBERG, 1965; WEBB, 1967; KWAN and WEBB, 1967). Reactions leading to the incorporation of

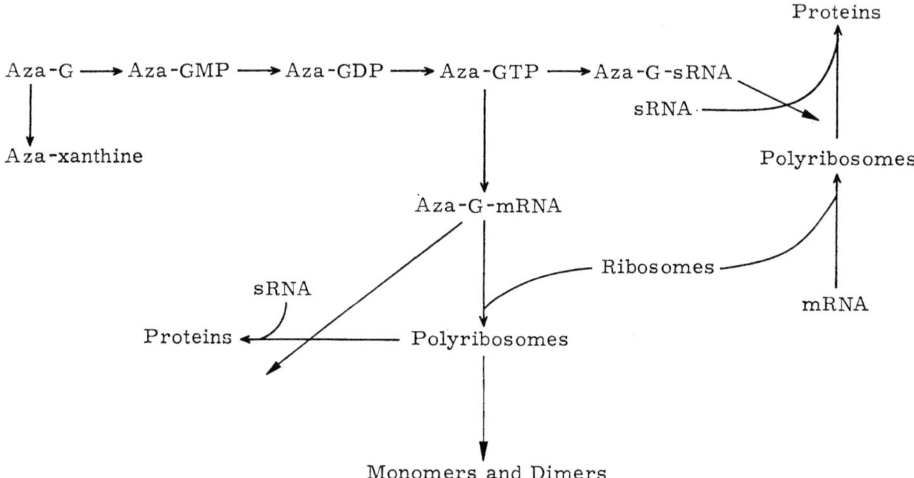

Fig. 16. Metabolism of aza-G and the possible inhibition sites. (The lines with ▶ indicate inhibitory sites)

aza-G into these species of RNA and the potential sites of inhibition are shown in Fig. 16.

The results of numerous studies have suggested that the growth inhibition produced by aza-G in bacterial, viral and mammalian systems is a consequence of the specific inhibition of protein biosynthesis. The inhibition of protein synthesis is related to the formaton of defective RNA since aza-G does not affect protein biosynthesis in reticulocytes (ZIMMERMAN and GREENBERG, 1965), which can convert the base to the ribonucleoside triphosphate (COOK and VIBERT, 1966) but which do not synthesize RNA. It has been observed in studies with *B. cereus* that the secondary structure of sRNA containing aza-G is less rigid than that of normal sRNA and as the percentage of guanine replacement by the analogue increases, the degree of order in the molecule decreases (LEVIN, 1963; LEVIN and LITT, 1964). However, the changes in secondary structure apparently do not inactivate the function of sRNA as aza-G-containing sRNA accepts amino acids (LEVIN, 1963) and the activity of this sRNA in the poly-uridylic acid directed polyphenylalanine synthesis remains quantitatively similar to those of normal sRNA (WEINSTEIN and GRUNBERGER, 1965). The results of more recent studies with intact or disrupted Hela cells (ZIMMERMAN and GREENBERG, 1965), *B. cereus* (ZIMMERMAN. HOLLER and PEARSON, 1967), and *E. coli* (GRUNBERGER et al., 1966) indicate that aza-G inhibits protein synthesis largely due to its incorporation into mRNA. WEBB (1967) has shown that the administration of 110 mg of aza-G per kg to rats results in the conversion of a fraction of the polyribosomes in regenerating liver and, to a lesser extent, in normal liver to monomers and dimers of ribosomes. Conditions which enhance the incorporation of the analogue into RNA are also found to increase breakdown of polyribosomes (KWAN and WEBB, 1967). Pulse labeling studies with radioactive amino acid have suggested that the monomers and dimers are inactive *in vivo*, while the residual polyribosomes in the livers of the treated animals incorporate the isotope at a near normal rate (KWAN and WEBB, 1967). The apparent selective effect of aza-G on certain polyribosomes and not on others may possibly be related to the half-lives of different mRNAs or to the location or amount of aza-G in those species of RNA (KWAN and WEBB, 1967).

From the above discussion it can be concluded that the inhibition of protein synthesis by aza-G is related to its incorporation into one or more species of RNA, thus rendering them defective in their functions. Preferential incorporation into one species of RNA and the resulting interference with protein synthesis may be depended on properties peculiar to the tissue concerned and the conditions involved (KWAN and WEBB, 1967). The inhibition of protein synthesis from chronic exposure of Hela cell cultures to low amounts of aza-G has been attributed to the production of fradulent mRNA, although only slight breakdown of polyribosomes takes place under these conditions (ZIMMERMAN and GREENBERG, 1965). Addition of polyribonucleotides to extracts from *B. cereus* treated with aza-G, causes a several fold increase in amino acid incorporation indicating that a large fraction of the ribosomes remains free of mRNA in the treated cells (ZIMMERMAN, HOLLER and PEARSON, 1967). Inhibition of protein synthesis in chick blastodiscs has been related to the sole incorporation of the base analogue into the sRNA fraction (WAINWRIGHT and WAINWRIGHT, 1967). Differences in template efficiency of codons in mRNA containing aza-G or guanine may also be a responsible factor. Aza-G can replace G in the GpUpU codeword for valine but the binding efficiency of the trinucleotide containing aza-G is lower than that of the normal triplet (GRUNBERGER et al., 1966). The template activity of aza-G containing triplets, aza-GpUpG, GpUpaza-G, and aza-GpUpaza-G, has been compared with that of GpUpG codon for valine (GRUNBERGER, HOLY and SORM, 1968). The substitution of one of the G residues by aza-G results in a lower template activity, and when both of the residues are replaced by aza-G the recognition by the triplet is highly diminished. A possibility has been considered that the growing polypeptide chains may terminate prematurely when the anomalous aza-GpUpaza-G sequence in the mRNA molecule is reached. This gains support from the observations that during the protein synthesis in the ribosomal system from aza-G-treated *B. cereus* cells most of the amino acids are incorporated into unfinished polypeptide chains which remain bound to the ribosomal particles and are not released into the supernatant fraction (GRUNBERGER, HOLY and SORM, 1968). An interesting possibility which has not yet been investigated is that the substitution of guanine by aza-G in the chain initiator codon, ApUpG, may interfere with the process of protein chain initiation or limit ribosome addition to mRNA.

In addition to the interference of aza-G on protein synthesis, several reports describe its effects on other enzymatic reactions. Its ribonucleoside triphosphate, aza-GTP, can function in place of GTP in the conversion of IMP to adenylosuccinate (SAMP) (reaction 7) catalyzed by adenylosuccinate synthetase (COHEN and PARKS,

$$IMP + aspartate + GTP \rightarrow SAMP + GDP + Pi \tag{7}$$

1963). The Michaelis constant and maximal velocity for the reaction obtained with the analogue are lower than those obtained with GTP. As a result, aza-GTP increases the reaction rate at low GTP concentrations and inhibits competitively at high GTP concentrations. However, this effect of aza-G can not be attributed to growth inhibition unless there is an accumulation of aza-GTP in high concentrations in specific tissues. Participation of aza-GDP as a substitute for GDP in reactions 8 and 9, feed-

$$succinate + CoA + GTP \rightleftharpoons succinyl\text{-}CoA + GDP + Pi \tag{8}$$

$$Phosphophenylpyruvate + CO_2 + GDP \rightleftharpoons Oxaloacetate + GTP \tag{9}$$

back inhibition of 5-phosphoribosylamine synthesis by aza-GMP, and formation of nicotinamide-azaguanine dinucleotide, have been reported by various workers (BROCKMAN and ANDERSON, 1963). The biological significance of aza-G inhibition of these sites will depend on the high levels of the aza-G nucleotides reached in growing cells, and as such these reactions may be considered to be of minor importance in the mechanism of growth inhibition by the analogue.

Resistance: Studies with microorganisms, mouse neoplasms *in vivo*, and mammalian cells in culture have shown that cellular resistance to aza-G is associated with decreased conversion of the analogue to the ribonucleotide due to partial or complete loss of guanine-hypoxanthine phosphoribosyltransferase activity. Since the catalyzing activity is believed to be due to a single enzyme acting on both guanine and hypoxanthine, its loss blocks the formation of GMP and its analogue as well as IMP in the salvage pathway. However, certain mutants of *Salmonella typhimurium* resistant to aza-G exhibit loss of guanine and not of hypoxanthine phosphoribosyltransferase activity (KALLE and GOTS, 1961). If a single enzyme is also responsible for the conversions in these mutants, it is attractive to speculate that enzyme may be modified during exposure to the inhibitor so as to escape its inhibitory effects. In other words it may be possible that the new enzyme formed in the resistant cells accepts only hypoxanthine as substrate and does not accept guanine or its congener (aza-G).

9-β-D-Arabinofuranosyladenine (Arabinosyladenine)

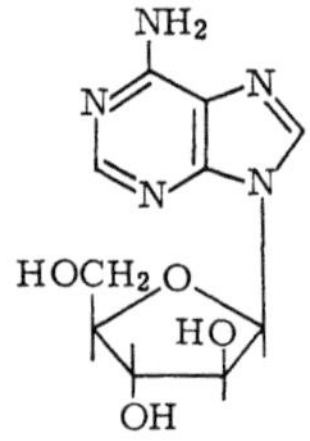

Arabinosyladenine 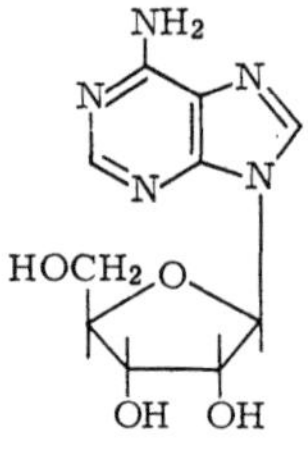   Adenosine 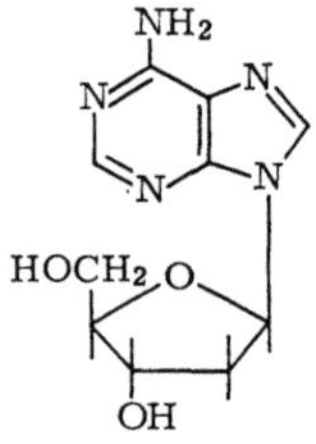   Deoxyadenosine

9-β-D-Arabinofuranosyladenine (arabinosyladenine or ara-A) may be considered to be a nucleoside analogue of either adenosine or deoxyadenosine in which the purine moiety remains unchanged but the sugar moiety is altered at the 2 position. This analogue was first synthesized by LEE et al. (1960) and its inhibitory effects on a purine-requiring strain of *E. coli* were studied by HUBERT-HABART and COHEN (1962). Subsequently it was shown that ara-A is effective in inhibiting the growth of a number of ascites tumors in mice and of mouse fibroblasts in suspended cell culture (BRINK and LePAGE, 1964; BRINK and LePAGE, 1964a; BRINK and LePAGE, 1965; DOERING, KELLER and COHEN, 1966). Its effectiveness in producing breaks in the chromosomes of human leukocytes induced to divide by phytohemaglutinin has been reported (NICHOLS, 1964). The antitumor effect of the analogue is decreased by its enzymatic deamination *in vivo* and the biologically inactive catabolite, arabinosyl-hypoxanthine is rapidly excreted (LePAGE and JUNGA, 1965). Thus, activity of ara-A as an inhibitor is related to the levels of adenosine deaminase in tumor tissues. A mouse tumor with relatively high adenosine deaminase activity, sarcoma 180, is practically unresponsive to ara-A while tumors with lower deaminase content, carcinoma 1025 and Ridgeway osteogenic sarcoma are responsive to the drug (KOSHIURA and LePAGE, 1968). Several nucleosides which are inhibitors of adenosine deaminase *in vitro* failed to potentiate the effect of ara-A *in vivo*. However, agents preventing entry of the deaminase substrates into blood cells showed some effectiveness. The most prolonged effects were obtained with 4-ethanolisopropanolamino-2,7-di (2'-methyl morpholino)-6-phenylpteridine (KOSHIURA and LePAGE, 1968).

In ascites cells ara-A is phosphorylated to the corresponding mono- (ara-AMP), di- (ara-ADP) and triphosphate (ara-ATP) derivatives (BRINK and LePAGE, 1964a). Evidence for the incorporation of ara-A into DNA of mammalian or bacterial cells could not be obtained (BRINK and LePAGE, 1964a; LePAGE and JUNGA, 1965). FURTH and COHEN (1967) observed that ara-ATP is inactive as a substrate for DNA polymerase or RNA polymerase isolated from either bacterial or mammalian sources. Although it was reported that ara-A is incorporated into RNA of ascites tumors to some extent (BRINK and LePAGE, 1964), the incorporation has been found to be insignificant in L cells or in *E. coli* cells (COHEN, 1966).

The possible reactions which are inhibited by ara-A nucleotides in mammalian cells are shown in Fig. 17. YORK and LePAGE (1966) studied the effect of ara-ATP on the incorporation of dTMP-^{14}C into DNA by DNA polymerase in cell-free extracts of TA3 ascites tumor cells. The results showed that ara-ATP is a non-competitive inhibitor of DNA polymerase and that an equimolar concentration of ara-ATP and the four common deoxyribonucleoside triphosphates produces a

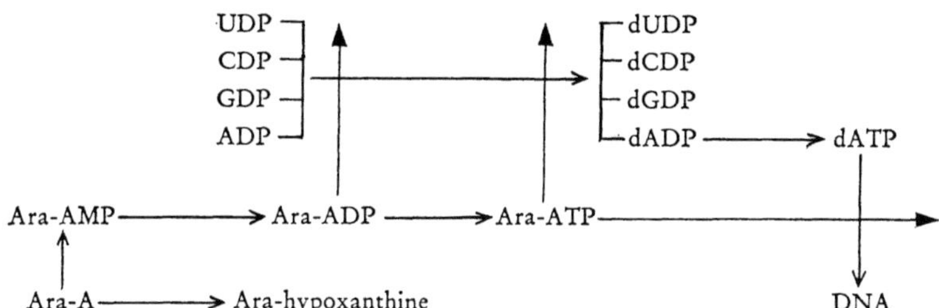

Fig. 17. Phosphorylation of ara-A to the active inhibitors and the possible *loci* of action in mammalian cells. Lines with ▶ indicate sites of inhibition

40—60% inhibition of DNA synthesis. The concentration of ara-ATP required for 40% inhibition of polymerase activity is roughly one-fourth the calculated value for the concentration of ara-ATP which accumulates inside TA3 ascites cells during inhibition by ara-A. These workers, therefore, stressed that the most probable effect of ara-ATP in the ascites cells is a direct interaction with the DNA polymerase to produce an inactive enzyme. Similarly, FURTH and COHEN (1967) have shown that the nucleoside triphosphate analogue inhibits DNA synthesis catalyzed by DNA polymerase isolated from calf thymus or bovine lymphosarcoma. The reaction kinetics suggest a form of mixed inhibition between ara-ATP and dATP. It is unlikely that the inhibition produced is due to incorporation of the arabinosylnucleotide into DNA, since under conditions in which 1.04 mμmoles of dTMP-^{3}H are incorporated into acid-insoluble form, less than 0.01 mμmole is incorporated with ara-ATP-^{3}H as substrate. From these results it is also apparent that if ara-ATP is utilized at all for DNA synthesis, it is utilized at less than one-hundredth the efficiency of dTTP. In contrast to the results with mammalian DNA polymerase, *E. coli* DNA polymerase is not significantly inhibited by ara-ATP. RNA polymerases of bacterial and mammalian sources are not inhibited by ara-ATP (FURTH and COHEN, 1967).

Although inhibition of DNA polymerase may be the mechanism by which ara-ATP inhibits DNA synthesis in cultured mammalian cells and tumors, this effect is not applicable to all systems susceptible to ara-A. For example, ara-A is known to inhibit DNA synthesis in *E. coli*. However, the low level of inhibition of *E. coli* DNA polymerase by ara-ATP suggests that DNA synthesis is inhibited by some other mechanism. As a matter of fact, the di- and triphosphates of ara-A have been shown to inhibit reduction of all four common ribonucleoside diphosphates to deoxyribonucleoside diphosphates with a partially purified enzyme system from rat tumor (MOORE and COHEN, 1967). The effects of ara-ADP and ara-ATP in this system appear to be quite comparable to those of dATP, except that dATP is five to eight times more active on a concentration basis. Interference with the reduction of CDP and possibly also of UDP is caused by the competition with ATP for the activation site. The effects on the reduction of purine nucleoside diphosphates are less clear. Some interference is produced by ara-ADP in the activation by ATP of the reduction of GTP, but neither substrate (ADP) nor activator (dGTP) is capable of reversing significantly the inhibition by ara-ATP of the reduction of ADP. In addition to these inhibitory actions, it has been reported that mammalian amino acid activating systems are inhibited by ara-ATP (FURTH and COHEN, 1967). Polymerizations of ADP and CDP to polyadenylate and polycytidylate by polynucleotide phosphorylase of *E. coli* have been observed to be inhibited by ara-ADP (LUCAS-LENARD and COHEN, 1966).

9-β-D-Xylofuranosyladenine (Xylosyladenine)

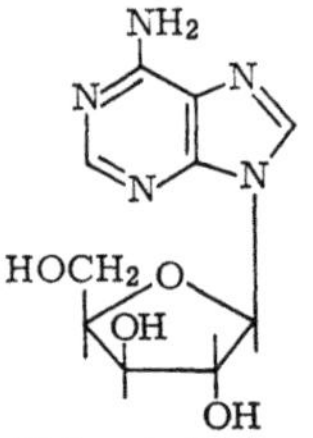

Xylosyladenine

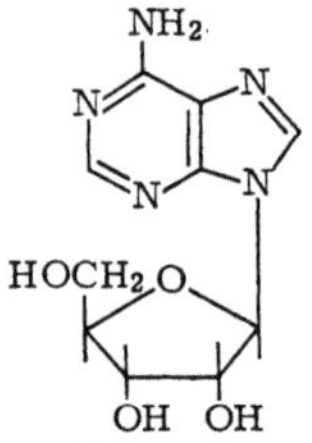

Adenosine

9-β-D-Xylofuranosyladenine (xylosyladenine or xyl-A) is an adenosine analogue in which the hydroxyl group on carbon 3 of the sugar is transposed. The nucleoside analogue was synthesized by CHANG and LYTHGOE (1950) and has shown effectiveness in prolonging the survival time of mice bearing TA3 or Ehrlich ascites tumors (ELLIS and LEPAGE, 1965). When administered to normal mice, xyl-A caused a loss in body weight without any effect on the blood leukocyte level or any other toxic manifestations (ELLIS and LEPAGE, 1966). In all tissues and tumors studied, relatively large amounts of the analogue are deaminated to xylosylhypoxanthine and rapidly eliminated from the cells in this form without any cleavage to free base (ELLIS and LEPAGE, 1966). A relationship between the levels of adenosine deaminase activity of neoplasms of experimental animals and humans and the effectiveness of xyl-A as an inhibitor has been observed (KOSHIURA and LEPAGE, 1968). Tumors with high adenosine deaminase activity showed little response to xyl-A while tumors with lower deaminase content were responsive to the drug.

As in the case of most of the purine analogues, xyl-A exerts its metabolic effects after conversion to the nucleotide form. It is rapidly phosphorylated to the triphosphate level by both normal and neoplastic tissues (ELLIS and LEPAGE, 1966). In ascites tumor cells *in vitro*, with an adequate glucose supply, the conversion of xyl-A to xyl-ATP is almost complete with little or no accumulation of xyl-ADP or xyl-AMP. Formation of xyl-ATP is less complete *in vivo*, where glucose supply may be expected to be limiting (ELLIS and LEPAGE, 1965). It is interesting that there is a marked difference in the ability of ascites cells to convert xyl-A and arabinosyladenine (ara-A) to the corresponding nucleoside triphosphate level. At identical levels of substrate and tumor cells, the amount of xyl-ATP formed is about twenty times higher than that of ara-ATP. The possibility that differences exist in the permeability of the cell to the two compounds has been eliminated and it has been suggested that these effects result from a difference in the *Km* values for the two substrates for adenosine kinase (YORK and LEPAGE, 1966).

The metabolic conversions and the known site of action of xyl-A are shown in Fig. 18. Accumulation of xyl-ATP in tumor cells causes an inhibition of the incorporation of ^{14}C-labeled adenine or glycine into both RNA and DNA, and into the acid-soluble ribonucleotide pool (ELLIS and LePAGE, 1965; ELLIS and LePAGE, 1965 a). It has been demonstrated that the inhibition of the conversion of adenine to its nucleotides by xyl-ATP in cell-free extracts of TA3 cells is the result of a decreased rate of PRPP formation form ATP and ribose 5-phosphate. When present in concentrations of 0.2, 0.4, and 0.8 mM, xyl-ATP inhibited nucleotide formation by 42,

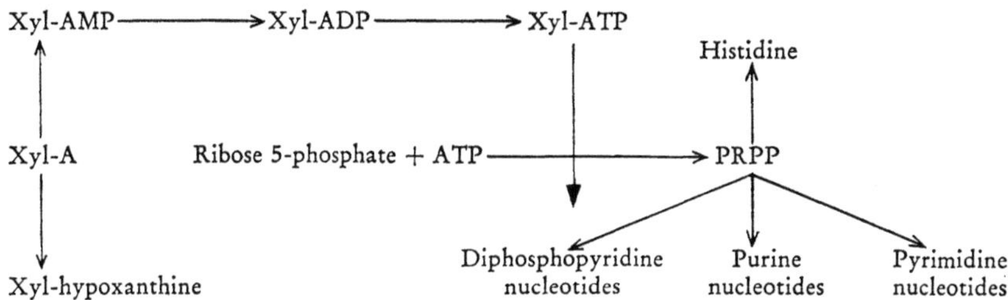

Fig. 18. The major inhibitory site (indicated by ▶) of the active metabolite of xyl-A

66, and 85%, while the nucleoside itself did not produce inhibitory effects even at a concentration of 2 mM (ELLIS and LePAGE, 1965 a). PRPP is a metabolite of major importance for purine and pyrimidine synthesis *de novo*, for tryptophan synthesis from anthranilic acid, for diphosphopyridine nucleotide synthesis and for histidine formation. Thus, interference by xyl-ATP with the formation of PRPP is likely to cause inhibition of cellular growth and multiplication.

In addition to the above major inhibitory site, there are other reactions which are known to be affected by xyl-A. It is as effective as adenosine as a feedback inhibitor of purine synthesis in mice by preventing 5′-phosphoribosyl-N-formylglycineamide formation. The feedback inhibition is probably not significant *in vivo* since adenosine cannot produce the increased survival times observed in mice treated with xyl-A. Xyl-ATP exerts a slight inhibitory effect on the synthesis of DNA in extracts of TA3 cells (YORK and LePAGE, 1966). The mechanism of action of xyl-A has not been thoroughly studied and it remains to be seen whether other important biological reactions requiring participation of ATP are inhibited by xyl-ATP.

Chapter 3

Pyrimidines

A. Metabolism of Pyrimidines

Prior to a discussion of the biochemical mechanism of action of pyrimidine analogues it is desirable to outline briefly the normal metabolic reactions of pyrimidines. The various aspects of the biosynthesis and interconversions of pyrimidine nucleotides have been considered by REICHARD (1959), CROSBIE (1960), ULBRICHT (1965), SUGINO (1965), GRAV (1967), and LARSSON and REICHARD (1967). The reviews by STADTMAN (1966), and BLAKLEY and VITOLS (1968) include the regulatory mechanisms involved in the control of pyrimidine metabolism.

The pathway for *de novo* pyrimidine synthesis as outlined in Fig. 19 appears to be operative in all organisms so far studied. The components of the pyrimidine ring are derived from ammonia, carbon dioxide and aspartic acid, and the heterocyclic ring is completed prior to conversion to a phosphorylated derivative. The first-step in this biosynthetic pathway is the formation of carbamylphosphate which is also a precursor for the biosynthesis of arginine. Uridine 5′-monophosphate (UMP) is the first essential pyrimidine nucleotide formed in the *de novo* pathway. Cytosine and thymine nucleotides originate from UMP through the series of reactions outlined in Fig. 20. The only known pathway for the formation of cytosine nucleotides is amination of UTP yielding CTP. Ammonia is the specific amino group donor in bacterial systems. However, glutamine is specifically required for CTP formation in homogenates from Novikoff hepatoma.

Pyrimidine deoxyribonucleotides are formed from the corresponding ribonucleotides. It should be noted that the ribonucleotide reductases of *E. coli* or animal tissues reduce ribonucleoside diphosphates to deoxyribonucleoside diphosphates, whereas the enzymes from *Lactobacillus leichmannii* require ribonucleoside triphosphates as substrates. In Fig. 20 the nucleoside diphosphates are shown as substrates for reduction since most of the studies which are included in the present discussion were carried out with *E. coli* or animal systems.

Formation of deoxythymidylic acid (dTMP) involves methylation of the pyrimidine ring. Then enzyme, deoxythymidylate synthetase (also called thymidylate synthetase) catalyzes the conversion of dUMP to dTMP with a concomitant oxidation of 5,10-methylene tetrahydrofolate ($CH_2 = FAH_4$) to dihydrofolate (FAH_2) (reaction 10). In this reaction the hydrogen atom at carbon 6 of $CH_2 = FAH_4$ is trans-

$$dUMP + CH_2 = FAH_4 \rightarrow dTMP + FAH_2 \tag{10}$$

ferred to the methylene carbon to form the methyl group of dTMP (see Fig. 30 under the discussion of 5-fluorouracil).

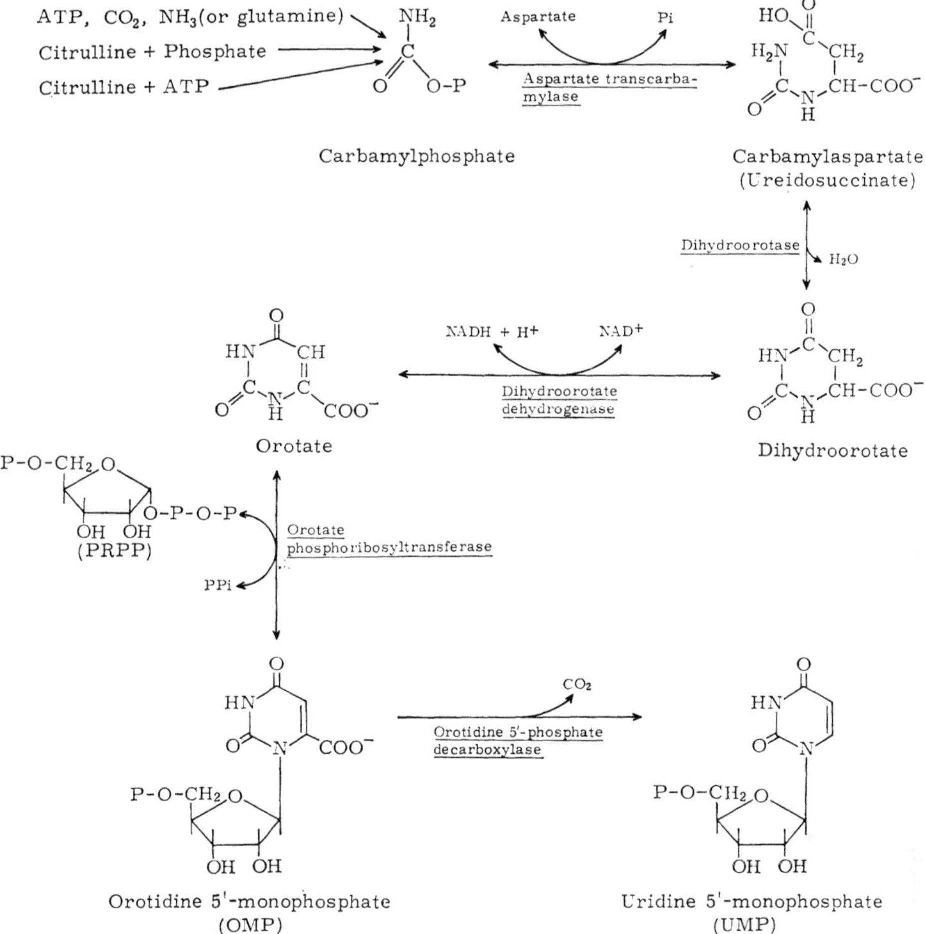

Fig. 19. The *de novo* pathway of uridine 5′-monophosphate biosynthesis

In addition to the *de novo* pathways, *salvage pathways* exist which utilize free bases arising from prior cleavage of nucleotides or exogenously supplied pyrimidines (Fig. 21 and 22). Uracil may be converted to UMP by reacting with 5-phosphoribosyl-1-pyrophosphate (PRPP) (Fig. 21). The enzyme catalyzing this reaction, uracil phosphoribosyltransferase is present in certain bacteria and animal tissues. Cytosine and thymine are not utilized by similar reactions. An alternate pathway exists in which the free base is converted to nucleotide *via* nucleoside (Fig. 22). The first step is catalyzed by nucleoside phosphorylases and the equillibrium of the reactions favors degradation. The specificity of nucleoside phosphorylases is not well documented. However, some nucleoside phosphorylases which act on uracil and thymine do not act on purines. The second step is catalyzed by nucleosidase kinase, and phosphate is donated by a ribo- or deoxyribonucleoside triphosphate. Several pyrimidine nucleoside kinases have been isolated which show specificity for nucleosides.

The pyrimidine deoxyribonucleoside triphosphates (dCTP, dTTP) and ribonucleoside triphosphates (UTP, CTP) are utilized for DNA and RNA synthesis,

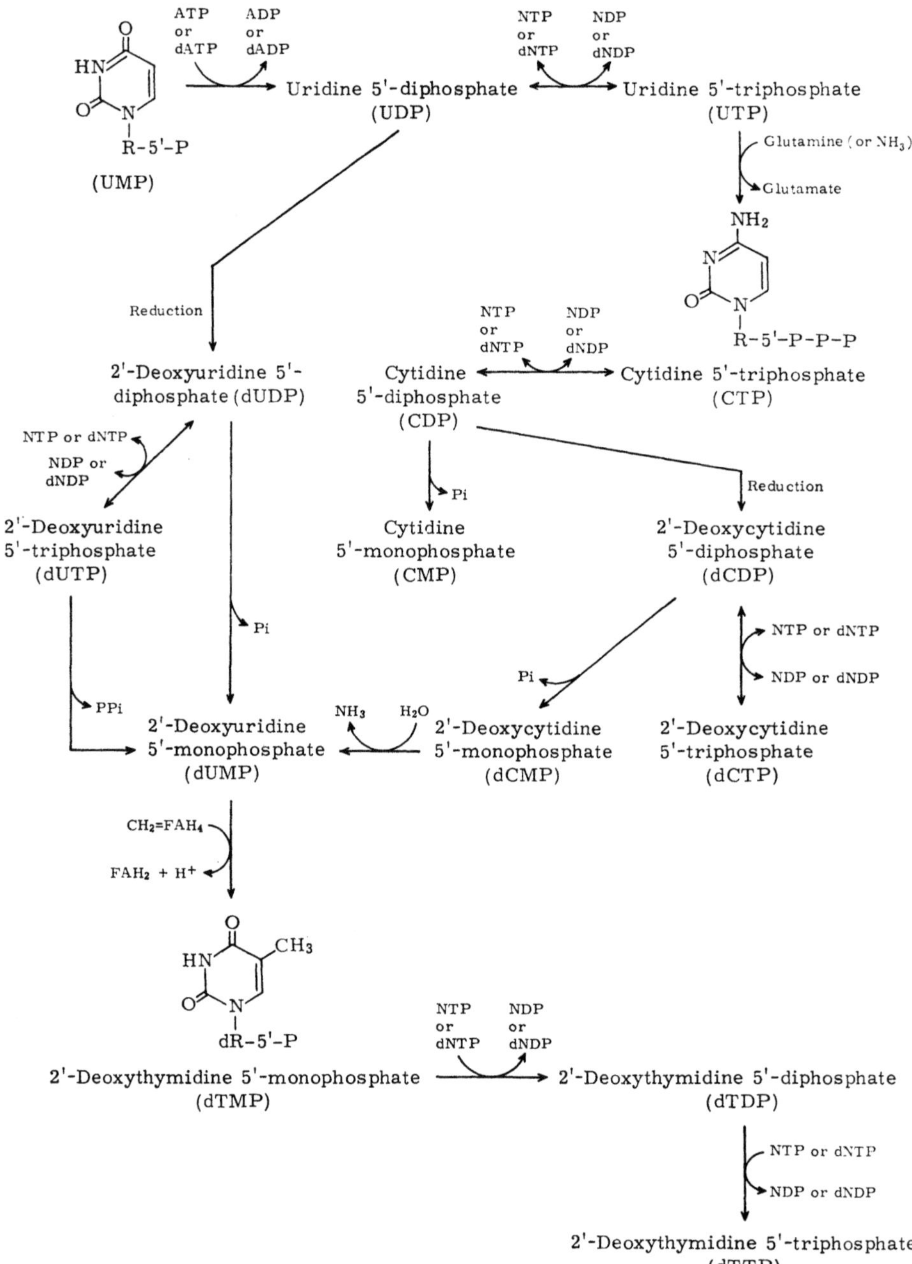

Fig. 20. Interconversions and phosphorylations of pyrimidine nucleotides (N = purine or pyrimidine ribonucleoside; d = deoxy; CH_2=FAH_4 = 5,10-methylene-tetrahydrofolate; FAH_2 = dihydrofolate)

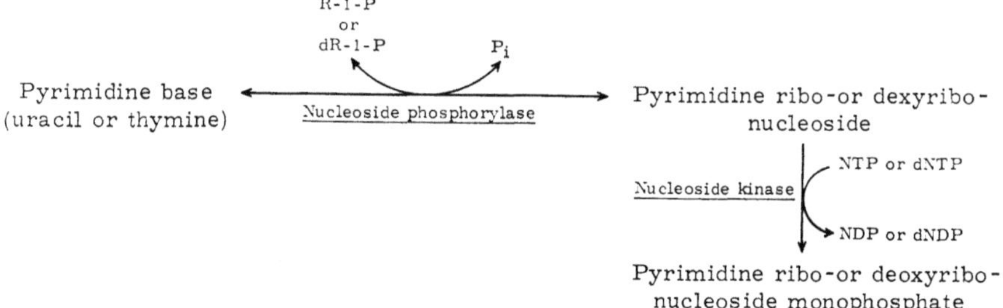

Fig. 21. Formation of uridine 5′-monophosphate directly from uracil

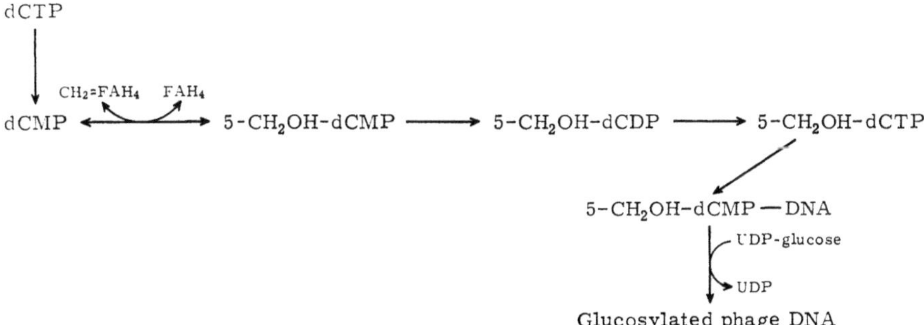

Fig. 22. Formation of pyrimidine nucleotides *via* pyrimidine nucleosides (N = purine or pyrimidine ribonucleoside; d = deoxy)

Fig. 23. Formation of glucosylated DNA in bacterial cells after T-even bacteriophage infection (ULBRICHT, 1965; YEH and GREENBERG, 1967)

respectively, as described in Chapter 2 (pages 10, 11). Deoxyribonucleic acids of T-even phages contain 5-hydroxymethylcytosine in place of cytosine. Bacterial cells infected with these phages produce an enzyme, deoxycytidylate hydroxymethylase, which catalyzes the addition of formaldehyde to deoxycytidine monophosphate at the C-5 position of the pyrimidine ring in the presence of tetrahydrofolate as coenzyme. Formation of this enzyme is induced after host infection only, and it is absent in uninfected cells. The DNA of T-even phages are partially or completely glucosylated at the hydroxymethylcytosine sites. Enzyme activities catalyzing transfer of glucose from UDP-glucose to hydroxymethylcytosine in DNA with the formation of α or β linkage have been demonstrated. Some of the metabolic changes following phage infection are outlined in Fig. 23.

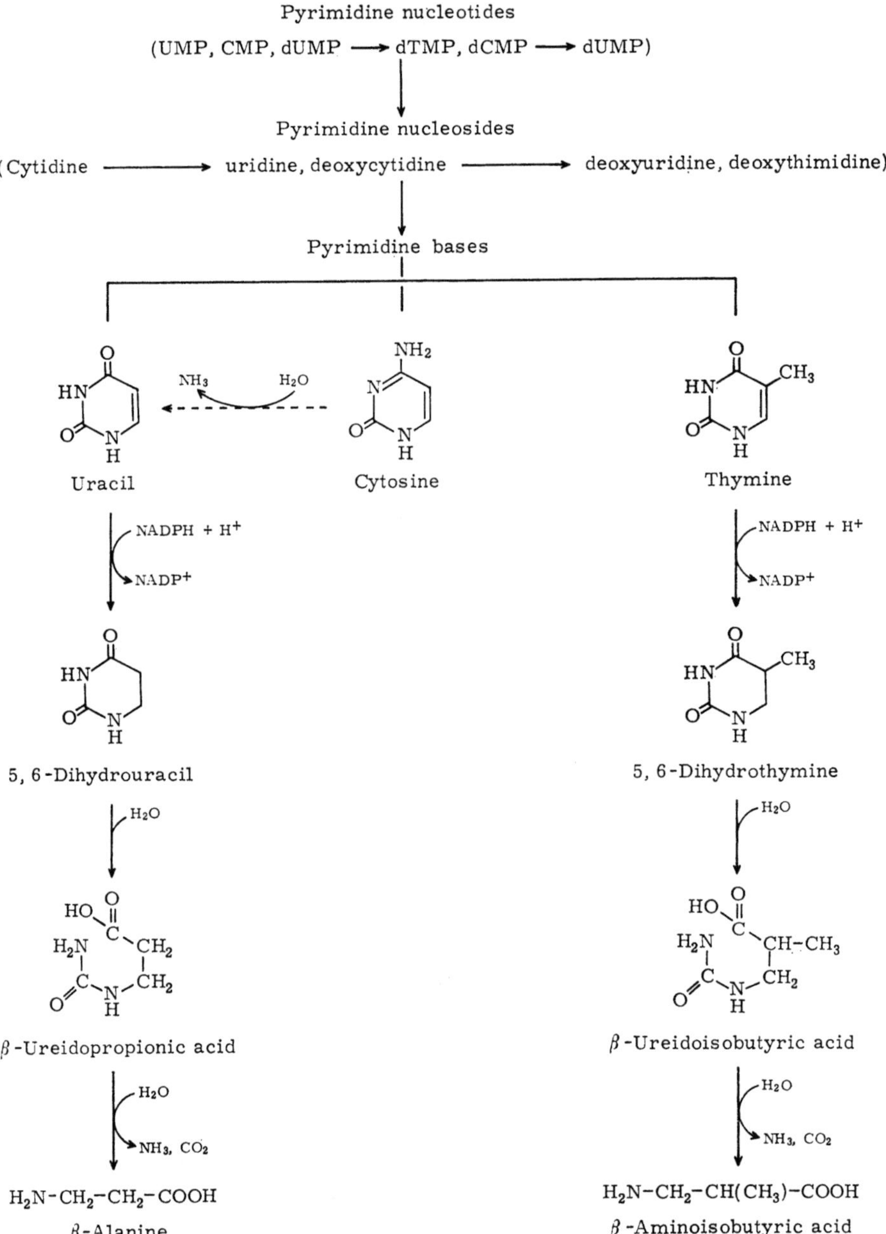

Fig. 24. Degradative reactions of pyrimidines

Pyrimidine nucleotide derivatives e. g., UDP-glucose, UDP-galactose, UDP-glucuronate, UDP-xylose, UDP-N-acetylmuramyl-pentapeptide, CDP-choline, TDP-rhamnose, participate in the biosynthesis of polyglycan, glycoproteins, phospholipids, monosaccharides and other important biological compounds. Some of these metabolites are formed by the action of pyrophosphorylases on pyrimidine nucleoside triphosphates in the presence of a phosphorylated substrate. For example, CDP-choline is formed from CTP and choline phosphate, and UDP-glucose from UTP and glucose-1-phosphate. Others are formed by enzymic interconversions of preformed derivatives.

In addition to the enzymes which catalyze the formation of nucleotides and poly-nucleotides there are several degradative enzymes operating at almost all levels of the internucleotide pathways. In the catabolic process, DNA and RNA are degraded by ribonucleases and deoxyribonucleases, of which there are many types, to ultimate and respective products of deoxyribo- and ribonucleoside monophosphates. The nucleoside monophosphates are dephosphorylated to nucleosides by various specific and non-specific phosphatases. The nucleosides are then split to free bases, presumably by the action of nucleoside phosphorylases. The stages in the breakdown of pyrimidines are shown in Fig. 24. The catabolism is carried out exclusively through uracil and thy-mine. Cytosine derivatives are degraded *via* uracil derivatives by enzymatic deamina-tion of dCMP to dUMP, deoxycytidine to deoxyuridine, or cytidine to uridine. In mammals uracil and thymine undergo a reduction step with the formation of 5,6-dihydro derivatives. This is followed by ring opening to ureido derivatives. Sub-sequent cleavages of these products give rise to β-amino acids which may enter the citric acid cycle.

As discussed in Chapter 2 (page 15) it is appropriate to mention at this point some regulatory reactions in which pyrimidine nucleotides may participate. These involve regulation of internucleotide metabolism by feedback controls. A summary of these regulations is outlined in Fig. 25. Bacterial aspartate transcarbamylase, catalyzing an early step in the *de novo* synthesis of pyrimidine nucleotides, is susceptible to allosteric inhibition by CTP and activation by ATP. However, in *Neurospora crassa* the

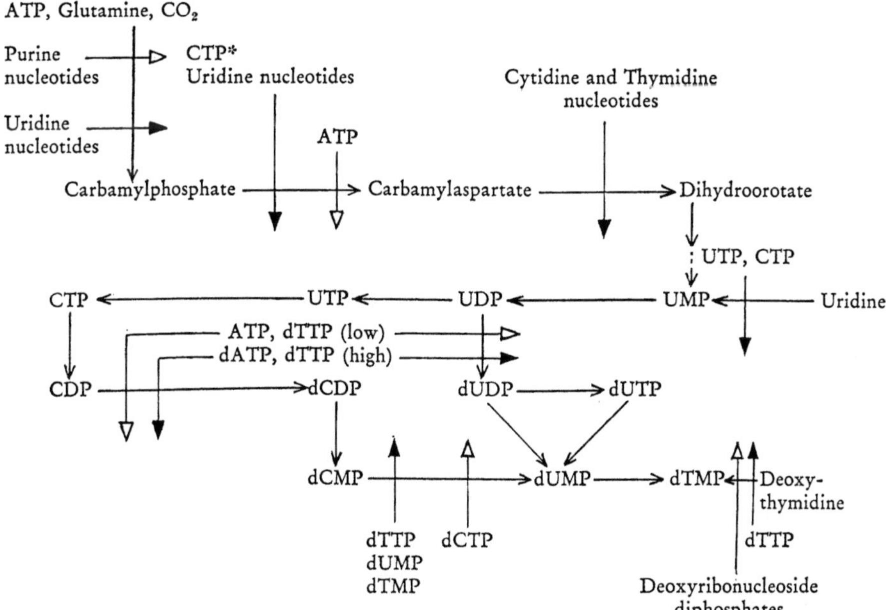

Fig. 25. Feedback controls in pyrimidine metabolism. (Thin lines indicate the steps that are subject to such controls, ▶ inhibition, ▷ stimulation; ∗, in bacterial systems)

enzyme is more effectively inhibited by UTP and other uridine derivatives, thymidine, and orotidylate than by cytidine derivatives. The enzyme of animal origin is also more sensitive to inhibition by uridine nucleotides. The interactions of the effector molecules with purified aspartate transcarbamylase from E. coli has been extensively studied. It appears that here are two catalytic subnits and four regulatory subunits per molecule of the native enzyme. Since the catalytic subunits are structurally different from the regulatory subunits, the catalytically active sites are likely to be distinct from the allosteric binding sites where the effector molecules interact. Another allosteric control by which the level of deoxyribonucleotides for DNA synthesis is maintained appears to be of considerable interest. Reduction of the two pyrimidine ribonucleoside diphosphates (UDP and CDP) to the corresponding deoxyribonucleoside diphosphates is activated by ATP. One of the products, dUDP is converted to dTTP through a sequence of reactions. The initial concentrations of dTTP stimulates the reduction of UDP or CDP. However, an increase in its concentration results in an inhibition of the reactions. It is also interesting to note that the activity of carbamylphosphate synthetase, the first enzyme in the *de novo* pathway, is significantly affected by a number of purine and pyrimidine nucleotides (ANDERSON and MEISER, 1966). The activity of this enzyme from *E. coli* is inhibited by uridine nucleotides and stimulated by purine nucleotides and these effects are exerted maximally by the first nucleotides synthesized in each pathway and decrease as the number of steps required to synthesize the various nucleotides from IMP or UMP, respectively, increases.

B. Mechanisms of Action of Pyrimidine Analogues

6-Azauridine

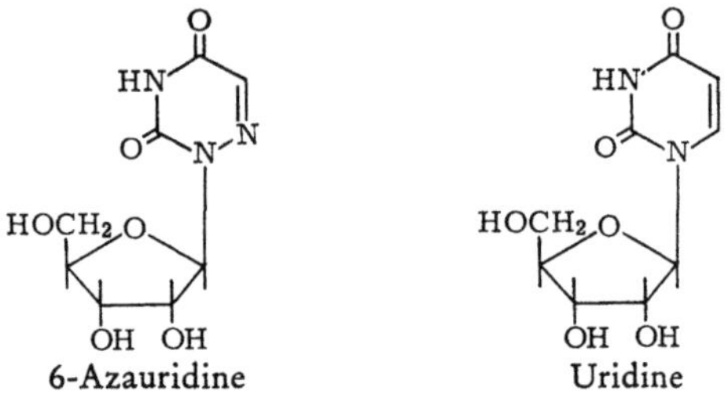

6-Azauridine   Uridine

The 1,2,4-triazine analogue of uracil, 6-azauracil, was first synthesized by SEIBERT in 1947. Biological studies with the analogue showed that it possesses bacteriostatic activity (SORM and SKODA, 1956; HANDSCHUMACHER and WELCH, 1956) and inhibits growth of a number of transplantable tumors (SORM, JAKUBOVIC and SLECHTA, 1956; BIEBER et al., 1957; JAFFE, HANDSCHUMACHER and WELCH, 1957; ELION et al., 1958). However, 6-azauracil was found to be toxic to the central nervous system and for this reason it is not clinically useful (WELCH, HANDSCHUMACHER and JAFFE, 1960; CALABRESI, 1963).

6-Azauridine (or 6-aza-UR) has been prepared by chemical and microbiological ribosylation of 6-azauracil (SCHINDLER and WELCH, 1957; SKODA, HESS and SORM, 1957; HANDSCHUMACHER, 1960). This ribosyl derivative has also demonstrated potent antineoplastic effects against a variety of transplantable tumors of rodents (JAFFE, HANDSCHUMACHER and WELCH, 1957; SORM and KEILOVA, 1958; CALABRESI and WELCH, 1965; CONN, CREASEY and CALABRESI, 1967). In contrast to the effects of 6-azauracil, intravenous administration of 6-aza-UR or its triacetyl derivative in man showed virtual absence of toxicity and indicated some initial beneficial responses in leukemia and some hyperplastic diseases (CARDOSO, CALABRESI and HANDSCHUMACHER, 1961; FALLON, FREI and FREIREICH, 1962; HANDSCHUMACHER et al., 1962; CREASEY et al., 1963; CALABRESI and TURNER, 1966). The triacetyl derivative of 6-aza-UR is considered to be an improved form of the drug because upon oral administration it is absorbed from the gastrointestinal tract without the toxicity that is associated with similarly administered 6-aza-UR which is poorly absorbed and is broken down to 6-azauracil by the intestinal flora (ELION and HITCHINGS, 1965).

6-Aza-UR is formed in bacteria grown in the presence of sub-bacteriostatic concentrations of 6-azauracil and the amounts formed are directly proportional to the amount of 6-azauracil added to the medium at the beginning of the logarithmic phase of growth (SKODA, 1963). This conversion is catalyzed by uridine phosphorylase in the presence of ribose-1-phosphate. In cell free extracts of *E. coli*, and in cells of lymphoma L1210 and Ehrlich ascites tumor 6-aza-UR has been shown to be phosphorylated to 6-aza-UMP (HANDSCHUMACHER and PASTERNAK, 1958; SKODA and SORM, 1959, SKODA, 1963). The enzyme, uridine kinase has been shown to participate in this phosphorylation reaction (SKODA and SORM, 1958, 1959; SKOLD, 1960). Further phosphorylation of 6-aza-UMP to di- and triphosphate forms has not been observed in mammalian cells (BROCKMAN and ANDERSON, 1963), although formation of these derivatives has been reported in microorganisms and protozoa (HANDSCHUMACHER, 1960; RUBIN, JAFFE and HANDSCHUMACHER, 1962). With isolated enzyme preparations of yeast it has been found that 6-aza-UDP is phosphorylated to 6-aza-UTP by nucleoside diphosphokinase but the product is inactive as a substrate for UDP-glucose pyrophosphorylase, the enzyme catalyzing the conversion of UTP and glucose-1-phosphate to UDP-glucose and pyrophosphate (ROY-BURMAN, 1969). 6-Aza-UTP is neither a substrate nor an inhibitor of RNA synthesis by *E. coli* RNA polymerase (KAHAN and HURWITZ, 1962). Most of the studies carried out with 6-aza-UR show that incorporation of the analogue into nucleic acid does not occur except in a few instances where some minor incorporation into RNA has been observed. HANDSCHUMACHER (1960) reported some labeling of RNA from radio-active 6-aza-UR in *Streptococcus faecalis*, and PINSKY and KROOTH (1967) observed that a minute amount of the label from 6-aza-UR-³H is incorporated into the acid-insoluble material of a mutant human diploid cells. The metabolism of 6-aza-UR in bacterial and mammalian cells and its major site of inhibitory action are outlined in Fig. 26.

The growth inhibitory effects of 6-aza-UR may be attributed mainly to its suppressive action of the *de novo* pathway of pyrimidine biosynthesis. At the enzymatic level this inhibition is produced by the interaction of orotidylate decarboxylase of bacterial or mammalian origin with the phosphorylated derivative of the analogue, 6-aza-UMP (HANDSCHUMACHER and PASTERNAK, 1958; PASTERNAK and HANDSCHUMACHER, 1959). In accordance with this finding the antitumor effect of 6-aza-UR has been found to be related to inhibition of orotic acid metabolism *in vitro* with a series of murine ascites and plasma cell tumors (BRUEMMER, HOLLAND and SHEEHE, 1962; CONN, CREASEY and CALABRESI, 1967). It is expected that inhibition of orotidylate decarboxylase would result in accumulation of orotidylate, its precursors and its degraded products of orotidine and orotic acid. This explains the large amounts of orotic acid and orotidylate accumulation in bacterial cultures under the influence of the inhibitor, and the excessive excretion of orotic acid and orotidine in animal and man after administration of the inhibitor (SKODA, 1963). An interesting observation related to accumulation of precursors has been made by PINSKY and KROOTH (1967).

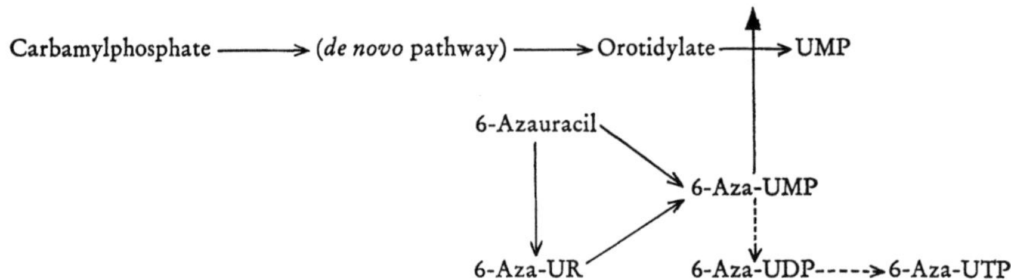

Fig. 26. Metabolism of 6-Aza-UR and its major site of inhibition (as indicated by ▶). The reactions shown in broken lines are known to occur in bacterial and not in mammalian systems

They found that cultured human diploid cell strains grown in the presence of 6-aza-UR develop augmented levels of both orotic acid phosphoribosyltransferase and orotidylate decarboxylase. A cell strain from a patient with orotic aciduria, a disease associated with deficient activity of the above two enzymes, was also found to develop near normal enzyme activities when grown in 6-aza-UR. It was demonstrated that the analogue (after conversion to 6-aza-UMP) reversibly inhibits orotidylate decarboxylase *in situ* and that the augmented activity is due to increased enzyme within the cells. The mechanism is not fully understood but the results suggest that the cells respond to an accumulated level of one or more intermediates following the inhibition of orotidylate decarboxylase. It is possible that an active precursor is dihydroorotic acid (PINSKY and KROOTH, 1967a).

In addition to the major blockade of orotidylate decarboxylation, a few other reactions have been reported to be affected by 6-aza-UR or its metabolites. 6-aza-UR competes with uridine for uridine kinase in cell-free extracts of *E. coli*, but it is probably a minor site of inhibition since the analogue itself is converted to the nucleotide by this enzyme (SKODA, 1963). KALOUSEK, RYCHLIK and SORM (1962) reported that in cell-free bacterial systems 6-aza-UDP interferes with the amino-cylation of transfer RNA and suggested that the effect is probably related to blocking of the formation of the terminal pCpCpA sequence. While 6-azauracil, 6-aza-UR and 6-aza-UMP all are essentially inactive in this respect, the results do not ascertain whether 6-aza-UDP or its further phosphorylated derivative 6-aza-UTP, is responsible for this effect. However, this *in vitro* observation is in disagreement with the findings that 6-aza-UR does not produce a detectable inhibition of general protein synthesis in whole cells (BROCKMAN and ANDERSON, 1963). Polynucleotide phosphorylase of *E. coli* is inhibited by 6-aza-UDP although the derivative is not a substrate for the enzyme (BROCKMAN and ANDERSON, 1963). It has also been suggested that the presence of the analogue may cause inhibition of synthesis of certain RNA species in bacteria (HABERMAN, 1961) and GOLDBERG and RABINOWITZ (1963) reported that 6-aza-UTP acts as an inhibitor of *E. coli* RNA polymerase. However, the concentration of the analogue triphosphate required to produce inhibition is much higher than that needed for orotidylate decarboxylase inhibition by 6-aza-UMP and moreover, KAHAN and HURWITZ (1962) did not observe a significant inhibition of the same polymerase by 6-aza-UTP. Thus, it appears that in systems where 6-aza-UMP undergoes further phosphorylations, the products formed may not add significantly to other inhibitory effects of the analogue.

Resistance: From the foregoing discussion it may be concluded that 6-aza-UR is active only after conversion to the nucleoside monophosphate. The development of resistance in a 6-aza-UR-resistant line of L5178Y mouse lymphoma has been attributed to decreased activity of uridine kinase and in a 6-azauracil-resistant mutant of *Streptococcus faecalis* to decreased uridine phosphorylase activity (BROCKMAN and ANDERSON, 1963). It has also been suggested that an adaptive increase in the *de novo* pyrimidine synthesis may be related to some cases of resistance encountered clinically (FALLON, FREI and FREIREICH, 1962; HANDSCHUMACHER et al., 1962). In experiments with plasma cell tumors of mice, administration of the analogue was found to produce an early decrease in orotic acid utilization, followed by a rapid return to normal levels despite continued administration of the inhibitor (CONN, CREASEY and CALABRESI, 1967). However, when 6-aza-UR administration is discontinued, orotidylate decarboxylase activity immediately surpasses pretreatment levels and inhibition by the analogue, as assayed *in vitro,* rapidly reappears.

5-Azacytidine

5-Azacytidine (or 5-aza-CR), an analogue of cytidine, was synthesized by PISKALA and SORM in 1964. Its mode of action has been studied entirely by SORM and his research group. 5-Aza-CR has been shown to be a potent inhibitor of lymphoid leukemia in strain AK mice. Intraperitoneal administration of the analogue in a single dose of 100 mg/kg destroys nearly all the leukemic cells in fifty percent of the treated animals (SORM and VESELY, 1964). The compound also has shown powerful bacteriostatic activity (CIHAK and SORM, 1965 a). It is highly mutagenic, producing reversion of a proline auxotrophic strain of *E. coli* to prototrophy while 5-bromodeoxyuridine and 5-iododeoxyuridine are without any effect (FUCIK et al., 1965). The analogue is also known to inhibit production of viable phage particles in bacteria infected with bacteriophage T_4 or f_2 (DOSKOCIL and SORM, 1967; DOSKOCIL, PACES and SORM, 1967) and to inhibit growth of root meristem of *Vicia faba.* This latter effect is accompanied by chromosome stickiness and aberrations (FUCIK, SORMOVA and SORM, 1965).

Conversion of 5-aza-CR to its mono-, di-, and triphosphate derivatives in Ehrlich ascites cells has been reported by JUROVCIK et al. (1965). It is incorporated into RNA of Ehrlich ascites cells and *E. coli* (JUROVCIK et al., 1965; CIHAK, TYKVA and SORM, 1966; PACES, DOSKOCIL and SORM, 1968), and into DNA of *E. coli* (ZADRAZIL et al., 1965; PACES, DOSKOCIL and SORM, 1968). Metabolism of 5-aza-CR and its major inhibitory effects are shown in Fig. 27.

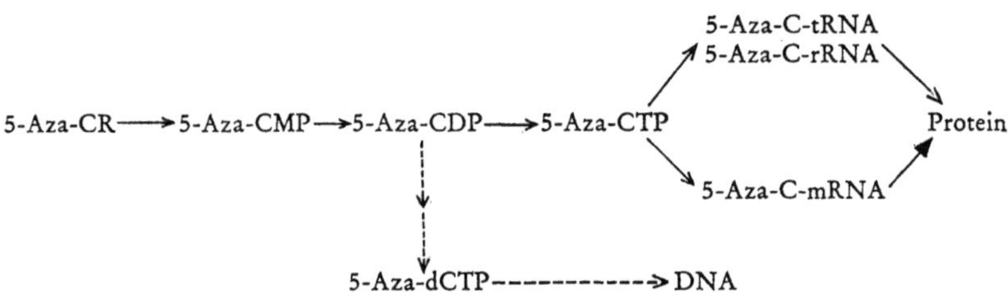

Fig. 27. Metabolism of 5-Aza-CR and its major site of inhibition (indicated by ▶). Broken lines represent reactions not definitely established

5-Aza-CR inhibits protein synthesis in various biological systems. This property has been considered to be the major reason for its inhibitory effect on growth. Two other effects of 5-aza-CR have been described. One is an inhibition of orotidylate decarboxylase by prior formation of 5-aza-CMP (VESELY, CIHAK and SORM, 1967) and the other is probable competition of the analogue or its nucleotides with normal substrates for kinases (RASKA et al., 1966).

The mechanism of inhibition of protein synthesis by 5-aza-CR has been investigated. RASKA et al. (1965, 1966) observed that the analogue is incorporated into isolated nuclei of calf thymus cells and inhibits incorporation of cytidine into nuclear RNA and of amino acids into nuclear protein. Induction of rat liver tryptophan pyrrolase is inhibited by 5-aza-CR and the inhibition produced is most pronounced when the compound is administered before or together with the inducer (CIHAK, VESELY and SORM, 1967). Administration after hormone treatment has a much smaller effect. The inhibitory effect of the analogue on this hormonal induction is supressed by simultaneous addition of a greater amount cytidine. On this basis and since the amount of incorporation of ^{14}C-orotic acid into liver RNA is very small as compared to the striking effect on induction, the authors suggest that the inhibition of the hormone-induced synthesis of the enzyme may be related to incorporation of 5-aza-CR into newly synthesized messenger RNA. Similarly, in *E. coli* it was observed that 5-aza-CR is capable of inhibiting leucine incorporation into total cell proteins as well as synthesis of two inducible enzymes, β-galactosidase and thymidine phosphorylase. Furthermore, during these inhibitions the rate of RNA synthesis is only slightly affected and DNA synthesis is not inhibited (DOSKOCIL, PACES and SORM, 1967). Recently, PACES, DOSKOCIL and SORM (1968) studied the kinetics of incorporation of ^{14}C-5-aza-CR into different nucleic acids in relationship to alterations in the rate of β-galactosidase synthesis in *E. coli*. They found that the analogue is incorporated into both DNA and RNA. A short exposure of the cells to ^{14}C-5-aza-CR followed by addition of cytidine results in transfer of radioactivity from rapidly labeled RNA to ribosomal and transfer RNA. All types of RNA containing 5-aza-CR have normal sedimentation constants and the pulse-labeled RNA is capable of forming hybrids with homologous DNA. Although there is no significant change in the physical properties of 5-aza-CR-containing RNA, the authors suggest that the synthesis of β-galactosidase is inhibited only when the analogue is present during the synthesis of mRNA. This is supported by the finding that when the incorporation of 5-aza-CR from the medium is prevented by addition of excess cytidine, resumption of protein synthesis occurs. The high turnover rate of the end sequence CpCpA of transfer RNA might lead one to expect that some of 5-aza-CR incorporated into tRNA would replace cytidine in this sequence. However, isolated tRNA from inhibited cells fails to show any significant variation from normal tRNA when tested for *in vitro* amino acid acceptor activity. Thus it appears that the primary inhibitory effect of 5-aza-CR on protein synthesis lies in its incorporation into mRNA. This view is supported by the observation that the analogue causes a strong inhibition of β-galactosidase induction when added simultaneously with the inducer, while addition of the inhibitor after the removal of the inducer has little effect (PACES, DOSKOCIL and SORM, 1968).

Its is inferred that incorporation of 5-aza-CR into mRNA interferes with the translation process. Since the symmetrical triazine molecule is unstable and is readily degraded (PITHOVA et al., 1965), it has been considered that a similar degradation may also take place in nucleic acids after incorporation of 5-aza-CR into polynucleotides. Decreased denaturation temperatures of RNA and DNA containing 5-aza-CR has been cited as evidence (CIHAK, TYKVA and SORM, 1966; ZADRAZIL et al., 1965). However, the recent results of PACES, DOSKOCIL and SORM (1968) do not support extensive breakdown of the analogue-containing RNA. Since almost all of the radioactivity of 5-aza-CR initially incorporated into rapidly labeled RNA of *E. coli* is subsequently found in ribosomal and transfer RNA, it appears that a reutilization of 5-aza-CR takes places after breakdown of the initial product. Moreover, enzymatic hydrolysis of isolated analogue-containing RNA revealed that a considerable percentage of 5-aza-CR remains intact in the macromolecule. The authors suggest that inhibition of protein synthesis is probably not related to alteration of 5-azacytosine moieties in messenger RNA and instead they emphasize that the inhibition is likely to be due to the presence of intact 5-azacytosine in this species of RNA. It is not clear at this point exactly how this analogue causes errors in the functional activity of mRNA.

The inhibitory process of 5-aza-CR on bacteria infected with bacteriophage appears very interesting. In *E. coli* infected with phage T_4 the inhibition of DNA synthesis is the only observable effect, whereas in uninfected cells the analogue does not inhibit DNA synthesis (DOSKOCIL and SORM, 1967). The inhibition by 5-aza-CR of DNA synthesis in infected cells is probably not due to inhibition of early enzyme synthesis because total protein synthesis remains unaffected until long after infection. Direct assay of one of the early enzymes, deoxycytidylate hydroxymethylase, showed that the presence of the analogue does not affect its production. The results seem to indicate that the analogue interferes specifically with viral DNA replication although it is not known at which step this occurs. Apparently the target site is one or more reactions specific for phage DNA replication since there is no observable inhibition of DNA synthesis in host cells. The authors propose that the most likely site may be the blocking of 5-hydroxymethyldeoxycytidylate formation. The analogue cannot be hydroxymethylated because of the replacement of the 5-carbon atom by nitrogen. It would be interesting to investigate this possibility at the appropriate enzymic levels.

Resistance: Biochemical changes in a strain of AKR leukemic mice resistant to 5-aza-CR treatment have been studied (VESELY et al., 1966; VESELY, CIHAK and

Sorm, 1967). It was observed that incorporation of orotic acid into livers of resistant mice is significantly lower than that in animals carrying the analogue-sensitive parent strain of leukemia. Incorporation of uridine, cytidine and 5-aza-CR into ribonucleic acids is also depressed in the resistant leukemic cells. Vesely, Cihak and Sorm (1967) have shown that uridine kinase activity of mouse leukemic cells gradually decreases as the development of resistance is gradually increased in mice. The gradual shortening of the survival time of treated animals and the depression of this enzyme activity showed a continuous character. However, the amount of depression of uridine kinase activity cannot alone account for high amounts of depression of 5-aza-CR incorporation into nucleic acids and thus indicates the existence of other mechanisms (Vesely, Cihak and Sorm, 1968). For example, an enhancement of nucleoside triphosphatase activity specific for pyrimidines has been observed in leukemic cells treated with 5-aza-CR (Vesely, Cihak and Sorm, 1968 a). It has been suggested that the lowered incorporation of 5-aza-CR into nucleic acids of the analogue-resistant AKR cells may be related to diminished phosphorylation as well as an enhanced breakdown of 5-aza-CTP formed in these cells (Vesely, Cihak and Sorm, 1968).

5-Azaorotic Acid

5-Azaorotic acid   Orotic acid

5-Azaorotic acid (or 5-aza-OA) is a symmetrical triazine derivative of orotic acid. The analogue has shown marked inhibitory effect on the growth of mammary adenocarcinoma of mouse at nontoxic doses (Elion et al., 1958) and preliminary clinical trials have indicated its possible use as an antitumor agent in the therapy of liver and kidney tumors (Granat et al., 1965). The compound is excreted mostly unchanged with small amounts of N-formylbiuret, a degradation product.

Metabolism of 5-aza-OA and its possible sites of inhibition are outlined in Fig. 28. 5-Aza-OA is known to interfere with the metabolism of tissues which actively utilize orotic acid. Inhibition of orotate phosphoribosyltransferase, the enzyme catalyzing

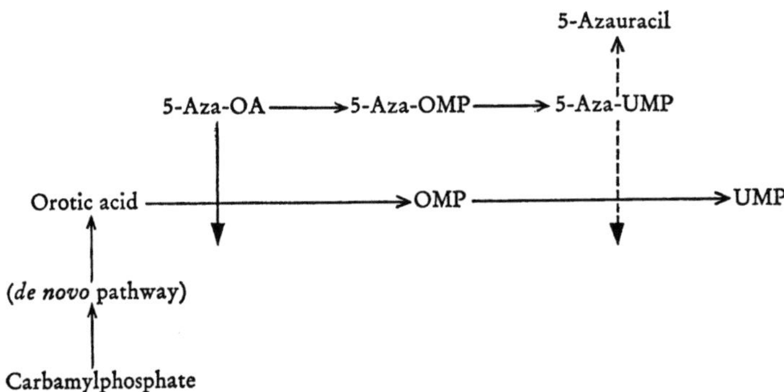

Fig. 28. Metabolic conversions of 5-Aza-OA and its possible *loci* of action. Lines with ▶ indicate sites of inhibition and the broken lines represent reactions not definitely established

the formation of orotidine 5'-monophosphate (OMP) from orotic acid and PRPP, by the analogue has been demonstrated in extracts from mouse liver, L5178Y leukemic cells, human diploid cell strains and in intact human cells (HANDSCHUMACHER, 1963; PINSKY and KROOTH, 1967; RUBIN, REYNARD and HANDSCHUMACHER, 1964). A 6×10^{-7} M concentration of 5-aza-OA was shown to produce approximately 50 percent inhibition of the metabolism of orotic acid in the extracts of mouse liver or leukemic cells (HANDSCHUMACHER, 1963). The inhibition of orotate phosphoribosyltransferase of intact mammalian cells by the analogue has been reported to be competitive (RUBIN, REYNARD and HANDSCHUMACHER, 1964).

CIHAK and SORM (1965, 1967) observed that in cell-free liver preparations and in the presence of PRPP 5-aza-OA is converted to 5-azaorotidine 5'-monophosphate (5-aza-OMP) and 5-azauridine 5'-monophosphate (5-aza-UMP) resembling the enzymic conversions of orotic acid. Formation of 5-azauracil in small amounts, probably by degradation of 5-aza-UMP, has also been observed. These metabolic transformations of 5-aza-OA indicated the possibility of additional sites of interference. Thus it was observed that addition of the analogue to liver extracts together with OMP does not affect the decarboxylation of OMP to UMP, but a short-term preincubation of liver extracts with 5-aza-OA produces a pronounced inhibition of OMP decarboxylase. Similarly, intraperitoneal administration of 5-aza-OA in mice at small doses causes a significant inhibition of the decarboxylase. On the basis of these results, CIHAK and SORM (1967) suggested that inhibition of OMP decarboxylase may be considered to be another site of action. It is very likely that the metabolic product, 5-aza-UMP is the active inhibitor of the enzyme.

5-Fluorouracil, 5-Fluorouridine, and 5-Fluoro-2'-deoxyuridine

5-Fluorouracil

Uracil

The uracil analogue, 5-fluorouracil (FU) was synthesized by DUSCHINSKY, PLEVEN and HEIDELBERGER (1957) and was first shown to have biological and tumor inhibitory activity by HEIDELBERGER et al. (1957, 1958). The analogue and its deoxyribonucleoside derivative, 5-fluoro-2'-deoxyuridine (FUdR) have been used clinically in the treatment of cancer and have produced objective responses in patients, particularly those with breast and gastrointestinal cancers (MONTGOMERY, 1965; KRAKOFF, 1967; REGELSON, 1967; AHMANN, BISEL and HAHN, 1967).

Fluoropyrimidines have been used in a variety of studies of their metabolism, of mechanisms of action and as tools for elucidating biochemical phenomena in a number of *in vitro* and *in vivo* biological systems. As a consequence the literature pertaining to these compounds is extensive. It is fortunate that the burden of this present discussion has been greatly reduced by the two recent reviews by HEIDELBERGER (1965, 1967) which have covered most of the field.

In mammalian and bacterial systems, FU is converted to its ribonucleoside (FUR) by uridine phosphorylase. The ribonucleoside is then phosphorylated to its nucleotide, FUMP by uridine kinase. Formation of FUMP may also result from direct conversion of FU to FUMP by uracil phosphoribosyltransferase. The nucleotide is further phosphorylated to FUDP, and is converted to its deoxyribonucleotide, dFUMP. The reaction sequence leading to the formation of dFUMP has not been investigated, however it is likely that dFUMP is formed by reduction of FUDP to dFUDP and subsequent dephosphorylation. Since the cellular occurrence of dFUDP or dFUTP has not been detected, it may be assumed that dephosphorylation of dFUDP to dFUMP is rapid and that dFUMP or dFUDP is not a substrate for the kinase which converts dUMP to dUDP or dUDP to dUTP. The deoxyribonucleoside FUdR is a good substrate for deoxythymidine kinase but the product dFUMP does not undergo further enzymatic phosphorylation (BRESNICK and THOMPSON, 1965). Flourouracil is incorporated *in vivo* into many ribonucleic acids including mammalian, bacterial, tobacco mosaic virus, polio virus and coliphage MS2 (HEIDELBERGER, 1965). In the cell-free *E. coli* RNA polymerase reaction it partially replaces UTP (KAHAN and HURWITZ, 1962). Some incorporation of the analogue into RNA when supplied as FUdR may be explained by the cleavage of the deoxyribonucleoside to FU which is then converted to FUTP by a sequence of reactions. Fluorouracil or FUdR is not incorporated into DNA of many organisms tested with the exception of *Bacillus subtilis* phage PBS2. The *in vitro* incorporation of dFUTP into DNA in reactions catalyzed by DNA polymerase (RICHARDSON, SCHILDKRAUT and KORNBERG, 1963) suggests that the *in vivo* exclusion of FU from incorporating into DNA may be related to inefficient production of dFUTP. Incorporation of FUdR into DNA of phage PBS2 may be considered as a special case because it is known that this phage DNA contains uracil in place of thymine. It has been shown that the uracil base of this DNA is specifically replaced by FU base and that such incorporation enhances the ultraviolet sensitivity of the phage (LOZERON and SZYBALSKI, 1967).

When degraded FU is converted to α-fluoro-β-ureidopropionic acid which is then broken down to urea, carbon dioxide, and α-fluoro-β-alanine (HEIDELBERGER, 1965). These degradative reactions are analogous to those of uracil and it can be assumed

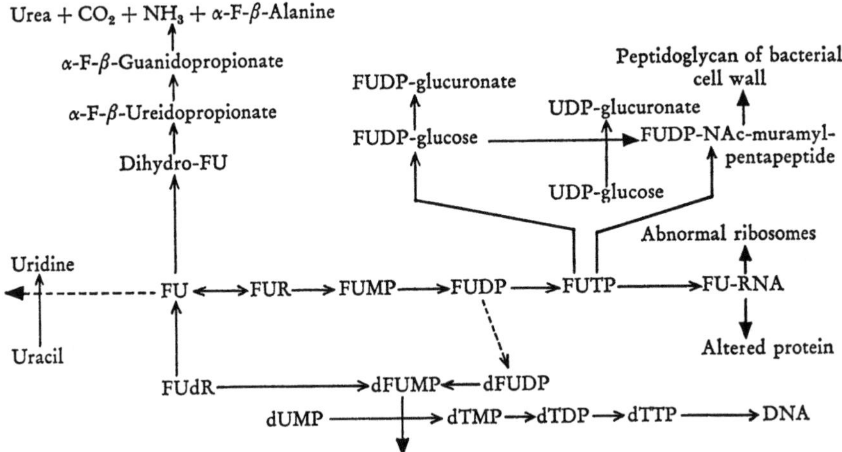

Fig. 29. Metabolism of FU, FUR and FUdR and their known *loci* of inhibition as indicated by ▶. (After HEIDELBERGER, 1965). Broken lines represent reactions not definitely established

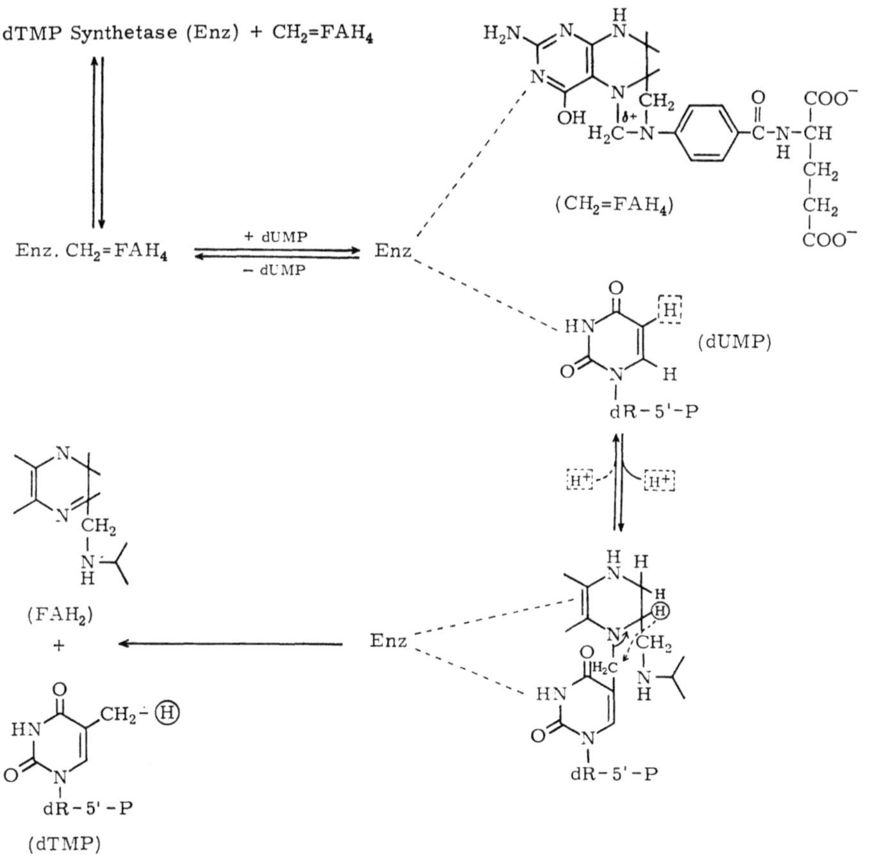

Fig. 30. Mechanism of dTMP synthetase reaction. $CH_2 = FAH_4 = $ 5,10-methylene-tetrahydrofolate, $FAH_2 = $ dihydrofolate (LOMAX and GREENBERG, 1967)

that they are catalyzed by the same enzymes. The degradation of FUdR follows the same course after its enzymatic cleavage to FU (BIRNIE, KROGER and HEIDELBERGER, 1963). Metabolism of FU, FUR and FUdR and the possible sites of inhibition are outlined in Fig. 29.

Several modes of inhibition have been proposed to explain the inhibitory effects of fluoropyrimidines on mammalian, bacterial or viral growths. For the present discussion these may be grouped as follows. (1) The deoxyribonucleotide (dFUMP) is a strong and specific inhibitor of deoxythymidylate (dTMP) synthetase which catalyzes the formation of dTMP from dUMP. This inhibition results in specific blocking of DNA synthesis. (2) Incorporation of FU into RNA in place of uracil may produce abnormal RNA species which in turn may cause aberrations in enzymes, proteins or in structural organizations of some biological particles. (3) Biosynthesis and utilization of uracil nucleotides may be inhibited by FU and its nucleotides resulting in a reduced rate of RNA or DNA synthesis or interference with the function of nucleotide sugar derivatives. It is possible that each of these effects contributes to the inhibition of growth. One or other of these effects may be more pronounced in one system than another. In the following paragraphs an attempt will be made to summarize the evidence favoring these inhibitory sites in the same sequence as they are mentioned above.

It would be appropriate to outline the mechanism of dTMP synthetase reaction prior to the discussion of its inhibition by dFUMP. This enzyme catalyzes the conversion of dUMP to dTMP with the concomitant oxidation of 5,10-methylene tetrahydrofolate ($CH_2 = FAH_4$) to dihydrofolate (FAH_2) as shown in reaction 10. The hydrogen on C-6 of $CH_2 = FAH_4$ is transferred to the methylene carbon to form the methyl group of dTMP (Fig. 30). The reaction proceeds in two steps. The first step is considered to be reversible as indicated by the hydrogen-exchange, while the second step is considered to be an irreversible rearrangement to form dTMP and FAH_2 (LOMAX and GREENBERG, 1967; LORENSON, MALEY and MALEY, 1967).

Inhibition of dTMP synthetase by dFUMP was first shown by COHEN et al. (1958) in bacteriophage-infected *E. coli*. It was shown that dFUMP is a potent competitive inhibitor of dUMP in this reaction. However, when the enzyme is incubated with the inhibitor prior to the addition of the substrate the inhibition appears noncompetitive (MATHEWS and COHEN, 1963). Similar noncompetitive kinetics were observed with a partially purified dTMP synthetase preparation from *Streptococcus faecalis* (BLAKELY, 1963). But with an enzyme preparation from Ehrlich ascites cells, HARTMAN and HEIDELBERGER (1961), and REYES and HEIDELBERGER (1965) obtained competitive inhibition whether or not the enzyme was preincubated with the inhibitor. In either case the inhibition produced is very efficient. The reported Km for dUMP is about 1.5×10^{-5} M whereas the Ki for dFUMP is reported to be 5.2×10^{-8} M without preincubation and 3.6×10^{-9} M with preincubation of the enzyme with the inhibitor (REYES and HEIDELBERGER, 1965). Recently, LORENSON, MALEY and MALEY (1967) observed that the purified synthetase from chick embryo extracts is inhibited in a competitive manner. However, the inhibition becomes noncompetitive when the enzyme is preincubated with the inhibitor in the absence of UMP. They also showed that the inhibition becomes competitive when the enzyme is incubated without $CH_2 = FAH_4$ and the reaction is initiated by the addition of the cofactor. This suggests that the cofactor is required for the altered binding of dFUMP and supports the dialysis experiments of REYES and HEIDELBERGER (1965) which demonstrated a marked enhancement of binding of dFUMP to the synthetase in the presence of $CH_2 = FAH_4$. The alteration in binding is further emphasized in the Ackermann-Potter plot (ACKERMANN and POTTER, 1949) showing a good stoichiometry in the binding of the inhibitor as a result of its prior incubation with the enzyme and cofactor (LORENSON, MALEY and MALEY, 1967).

Thus the mechanism of inhibition by dFUMP appears to be a complex process in the system studied. This cannot be explained by simple competition between dFUMP and dUMP for the enzyme site. The time-dependent alteration in inhibition kinetics from competitive to noncompetitive with chick embryo synthetase may be explained by a slow conformational change in the protein which is initiated by dFUMP in the presence of $CH_2 = FAH_4$. It is also possible that dFUMP may also bind to a second site on the enzyme which reacts more slowly than the first (LORENSON, MALEY and MALEY, 1967).

The inhibition of dTMP synthetase by dFUMP has been considered to be a major metabolic defect produced by FUdR, FUR or FU in mammalian systems (HEIDEL-

BERGER, 1967). Although this is probably true in most cases, there are notable exceptions. For example, a report by SOULINNA, SLAVIK and HAKALA (1967) compares the content of dTMP synthetase and the inhibition produced by FUdR in sarcoma 180 cells. It was found that the sensitivity of two different lines of cells to FUdR is nearly the same even though a 70-fold difference in the content of dTMP synthetase exists. Since the levels of nucleoside phosphorylase and deoxyuridine kinase are similar in both cell lines, the lack of correlation between the synthetase content and the degree of inhibition is not apparent.

Incorporation of FU into RNA has been considered to affect growth through the production of altered proteins (BUSSARD et al., 1960; GROS and NAONO, 1961; NAKADA and MAGASANIK, 1964). CHAMPE and BENZER (1962) showed that growth in the presence of FU restores the wild-type phenotype of certain mutants of bacteriophage T_4. Occurrence of such phenomena may result from base-pairing errors during the translation of FU-containing mRNA where FU may occasionally pair with guanine instead of adenine (CHAMPE and BENZER, 1962; ROSEN, 1965; EDLIN, 1965, 1965 a). In *E. coli* it has been shown that FU inhibits the synthesis of a number of enzymes, notably those of β-galactosidase (HOROWITZ, SAUKKONEN and CHARGAFF, 1959, 1960) and alkaline phosphatase (NAONO and GROS, 1960). It appears that an inactive protein immunologically related to β-galactosidase is formed in *E. coli* grown in the presence of the analogue (BUSSARD et al., 1960; NAKADA and MAGASANIK, 1964). There are two interesting effects of FU on viral replication which may be related in one case to a defect in the completion of viral RNA synthesis and in the second case to a possible defect in specific host protein synthesis. SHIMURA, MOSES and NATHANS (1967) observed that FU causes the formation of a class of FU-containing coliphage MS2 in *E. coli* with buoyant density lower than normal MS2. They are noninfectious and do not adsorb to *E.coli* receptor sites. Analysis of protein and RNA content of these altered phage particles shows that the RNA fragment lacks a specific part of the phage RNA molecule (SHIMURA, MOSES and NATHAN, 1967). On the other hand, such a direct effect of FU on the RNA of polyoma virus in mouse embryo cells is not apparent since the inhibition of viral multiplication occurs even when the analogue is not available in the intracellular pool for incorporation into viral mRNA (BEN-PORAT, KAPLAN and TENNANT, 1967). Since the degree of inhibition of this virus multiplication is dependent upon the length of time the cultures are exposed to FU prior to infection and since the infective process is suppressed in some of the cells and not in other cells, it appears that an early step in the infective process may be sensitive to FU.

The observations mentioned in the preceding paragraph indicate an inhibitory mechanism in which FU is incorporated into RNA to yield, in certain cases, functionally defective RNA molecules. However, there are several other reports which apparently do not support such a mechanism. BUJARD and HEIDELBERGER (1966) studied the base-pairing errors in polyribonucleotide synthesis *in vitro* and showed that less than 1 in 3000 cytosine residues is replaced by FU. LOWRIE and BERGQUIST (1968) reported that tRNA synthesized in the presence of FU may have up to 100 percent replacement of uracil by the analogue. Although this FU-containing transfer RNA possesses an altered secondary structure as judged by its thermal denaturation profile, it is capable of accepting and transferring amino acid in protein synthesis *in vitro*.

Tobacco mosaic and polio virus particles containing a significant percent of FU in place of uracil have essentially normal infectivity (GORDON and STAEHELIN, 1959; MUNYON and SALZMAN, 1962; HOLOUBEK, 1963) and it has been reported that functional mRNA specific for β-galactosidase can be synthesized by *E. coli* in the presence of FU (HOROWITZ and KOHLMEIER, 1967). In addition to these reports, there is evidence that extensive incorporation of FU into RNA or synthetic polyribonucleotides has little or no effect on the coding or other biological properties of these macromolecules (HEIDELBERGER, 1965).

The influence of FU on the structural organization of ribosomes has been studied recently following the observation of ARONSON (1961) that abnormal ribosomal particles are produced in *E. coli* cells grown in the presence of this analogue. These particles, termed FU-particles, are less stable than the normal ribosomal particles, both *in vivo* and *in vitro*, and their RNA moiety contains FU in place of about 70 percent of the uracil normally present (ANDOH and CHARGAFF, 1965). Sedimentation characteristics of FU-particles differ from those of normal ribosomes (OSAWA, 1965). For example, instead of the normal 30s and 50s ribosomal subunits a series of particles with sedimentation coefficients of approximately 27s, 33s, 40s and 47s is produced in *E. coli* (ANDOH and CHARGAFF, 1965). Blocking of normal ribosome biosynthesis by FU and formation of structurally different non-functional ribosomes have also been demonstrated in *Staphylococcus aureus* (strain DUNCAN), *Trichoderma* and in rat liver (HIGNETT, 1966; GRESSEL and GALUN, 1966; WILLEN and STRENRAM, 1967). In *Saccharomyces carlsbergensis* the analogue causes a severe inhibition of ribosome and ribosomal RNA formation, whereas the synthesis of soluble and messenger RNA continues to a considerable extent (KLOET and STRIJKERT, 1966). In *Trichoderma* it produces an additional RNA species with sedimentation value between 4s and 5s (GRESSEL and GALUN, 1966). It has been speculated that many of the differed properties of these abnormal ribonucleoprotein components in FU-treated cells could be due to alterations in the secondary structure of the RNA moieties resulting from the replacement of uracil by FU (ANDOH and CHARGAFF, 1965; HILLS and HOROWITZ, 1966). Changes in the secondary structure of the RNA components may interfere with the addition of protein to partially completed particles or the particles containing a full complement of protein may have to attain a different configuration in order to accommodate the altered RNA structure.

Pyrimidine nucleotide and nucleotide sugar metabolism may be affected by FU and its derivatives. FUTP can substitute for UTP as a substrate for UDP-glucose pyrophosphorylase (PARKS, WAY and DAHL, 1958). The product, FUDP-glucose is an alternate-substrate competitive inhibitor of UDP-glucose dehydrogenase (GOLDBERG, DAHL and PARKS, 1963) (reactions 11 and 12). The inhibition of bacterial cell wall

$$\text{UDP-glucose} + 2\,\text{NAD}^+ \xrightarrow{\text{UDP-glucose dehydrogenase}} \text{UDP-glucuronate} + 2\,\text{NADH} + 2\,\text{H}^+ \quad (11)$$

$$\text{FUDP-glucose} + 2\,\text{NAD}^+ \xrightarrow{\text{UDP-glucose dehydrogenase}} \text{FUDP-glucuronate} + 2\,\text{NADH} + 2\,\text{H}^+ \quad (12)$$

formation has been related to incorporation of FU into certain nucleotide sugar derivatives. Experiments of TOMASZ and BOREK (1959, 1960, 1962) showed that when *E. coli* K-12 is grown in the presence of FU, spheroplast-like bodies are formed which lyse in a medium of low osmotic strength. It was also observed that the analogue inhibits the incorporation of α,ε-diaminopimelic acid into the cell wall and that large amounts of N-acetylhexoseamine esters accumulate. ROGERS and PERKINS (1960) reported the accumulation of FUDP-N-acetylmuramyl-L-ala-D-glu-L-lys-D-ala-D-ala in *Staphylococcus aureus* grown in the presence of FU. Such accumulation results in an inhibition of the incorporation of amino acids into the peptidoglycan of the cell wall. The initial stage in the biosynthesis of peptidoglycan is catalyzed by phospho-N-acetylmuramyl-pentapeptide translocase (reaction 13). The rate of transfer of

$$\text{UDP-NAc-muramyl-pentapeptide} + \text{lipid-glycerol-phosphate} \rightleftharpoons$$
$$\text{lipid-glycerol-diphosphate-NAc-muramyl-pentapeptide} + \text{UMP} \qquad (13)$$

phospho-NAc-muramyl-pentapeptide from FUDP-NAc-muramyl-pentapeptide to lipid-glycerol-phosphate is about 2 percent of that observed with the UDP-derivative (STICKGOLD and NEUHAUS, 1967). The low activity of this FUDP-derivative for the translocase may account for its accumulation and for the inhibition of peptidoglycan synthesis. In addition to this, the FUDP-derivative is a competitive inhibitor in the transfer reaction from the normal substrate, the Ki and Km values being 1.2×10^{-4} M and 2.0×10^{-6} M respectively.

Resistance: The possible mechanisms for the development of resistance to FU, FUR and FUdR in bacteria, animals and human tumors have been reviewed and discussed by HEIDELBERGER (1965). Decreased cellular capacity to convert the analogue to its active phosphorylated forms has been considered to be a common mechanism. Thus, decreased uridine phosphorylase activity reduces the formation of FUR from FU. Lower uridine kinase levels results in lesser amounts of phosphorylation of FUR to FUMP. Reduced levels of deoxyuridine (thymidine) kinase decreases the formation of dFUMP from FUdR. A different mechanism may be operative in a particular resistant line of Ehrlich ascites tumor. It is possible that this resistant line has acquired an altered dTMP synthetase which is not inhibited by dFUMP.

5-Trifluoromethyl-2′-deoxyuridine ("Trifluorothymidine")

5-Trifluoromethyl-2′-deoxyuridine   Deoxythymidine

The nucleoside analogue, 5-trifluoromethyl-2′-deoxyuridine ("trifluorothymidine" or F_3TdR) was synthesized by HEIDELBERGER, PARSONS and REMY (1962, 1964). The van der Waals radius of the CF_3 group (2.44A) resembles the CH_3 group (2.00A) of thymine. This fluoropyrimidine derivative has a better therapeutic index than 5-fluoro-2′-deoxyuridine (FUdR) against adenocarcinoma 755 in mice (HEIDELBERGER and ANDERSON, 1964).

The analogue is phosphorylated readily by deoxythymidine kinase to its nucleotide, dF$_3$TMP, which is further phosphorylated by appropriate kinases to the di-(dF$_3$TDP) and triphosphate (dF$_3$TTP) derivatives (HEIDELBERGER, BOOHAR and KAMPSCHROER, 1965; BRESNICK and WILLIAMS, 1967). It is incorporated into DNA in variable amounts depending on the biological system. For example, 10.5 and 0.3 percent of the dTMP content of bacteriophage T$_4$ DNA, DNA of human bone marrow cells in culture, and DNA of adenocarcinoma 755 *in vivo*, respectively, is replaced by dF$_3$TMP (GOTTSCHLING and HEIDELBERGER, 1963; SZYBALSKI, COHN and HEIDELBERGER, 1963; HEIDELBERGER, BOOHAR and KAMPSCHROER, 1965). In the degradative reactions F$_3$TdR is converted *in vivo* to free base, trifluorothymine, and to a small extent to 5-carboxyuracil (HEIDELBERGER et al., 1963; HEIDELBERGER, BOOHAR and KAMPSCHROER, 1965). Metabolism of this analogue and the reactions which are known to be inhibited by it and its derivatives are shown in Fig. 31.

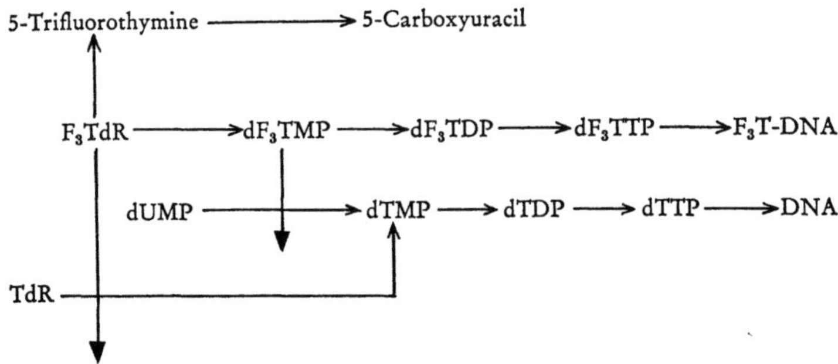

Fig. 31. Metabolism of F$_3$TdR and its known sites of inhibition (indicated by ▶)

Since F$_3$TdR acts as a substrate for deoxythymidine kinase the conversion of deoxythymidine to dTMP by the enzyme is inhibited in the presence of the analogue. The reported values for *Km* and *Ki* are 3.3×10^{-6} M and 3.7×10^{-6} M respectively, and the inhibition of dTMP synthesis appears as a complex function of ATP concentration (BRESNICK and WILLIAMS, 1967). The phosphorylated product of the analogue, dF$_3$TMP is a strong inhibitor of deoxythymidylate synthetase (reaction 10). The kinetics of inhibition of this enzyme from Ehrlich ascites cells show that dF$_3$TMP inhibits the synthetase competitively unless the enzyme is preincubated with dF$_3$TMP. In this case the inhibition is noncompetitive (REYES and HEIDELBERGER, 1965; HEIDELBERGER, 1965). The reported values for *Km* and *Ki* are 1.9×10^{-5} M and 3.6×10^{-8} M respectively. The ability of trifluorothymine moiety to undergo hydrolysis under very mild alkaline conditions to produce 5-carboxyuracil may be invoked to explain the time dependent alteration of the inhibition kinetics by a possible alkylation of an amino group at the active site of the enzyme (HEIDELBERGER, PARSONS and REMY, 1964; HEIDELBERGER, 1965). In addition to the interferences on deoxythymidine kinase and dTMP synthetase, there are various other potential sites of inhibition which have not been investigated. Since F$_3$TdR is incorporated into DNA it may be expected that such incorporation under certain cases could cause abnormalities in replication or transcription process.

5-Iodo-2′-deoxyuridine (Iododeoxyuridine)

Iododeoxyuridine   Deoxythymidine

The pyrimidine nucleoside analogue, 5-iodo-2′-deoxyuridine (IUdR) was prepared by iodination of deoxyuridine (PRUSOFF, 1959). The van der Waals radius of iodine atom (2.15A) resembles that of methyl group (2.0A). The compound has shown some antineoplastic activity in experimental tumors and in human subjects with advanced neoplastic disease, but leucopenia, stomatitis and alopecia are frequent toxic side effects in man (WELCH and PRUSOFF, 1960; CALABRESI et al., 1961; CALABRESI, 1963). Although IUdR is not considered as a useful agent in the treatment of neoplasms, its antiviral activity against some DNA virus infections in cell cultures, in animals, and in man has been established (PRUSOFF, 1963). Since the original observations of KAUFMAN and his collaborators (KAUFMAN, 1962; KAUFMAN, MARTOLA and DOHLMAN, 1962; KAUFMAN, NESBURN and MALONEY, 1962), numerous reports have appeared confirming the efficiency of IUdR in the treatment of herpes simplex virus infections of the corneal epithelium in man. These and other clinical results in the chemotherapy of viral diseases by IUdR have been reviewed by PRUSOFF (1967). This analogue also possesses radiation-sensitizing properties (ERIKSON and SZYBALSKI, 1963; FOX and LAJTHA, 1967).

IUdR is phosphorylated by deoxythymidine kinase to its monophosphate derivative, dIUMP which is further phosphorylated to dIUDP and dIUTP by the action of appropriate kinases. It may be incorporated into DNA to variable extent depending on the biological system (PRUSOFF, BAKHLE and SEKELY, 1965; PRUSOFF, 1967). In DNA viruses IUdR is incorporated readily in place of deoxythymidine. For example, this replacement is about 90 percent in pseudorabies viral DNA grown in rabbit kidney cells (KAPLAN and BEN-PORAT, 1966). In animal tissues, however, IUdR is less readily incorporated into DNA and the ratio between the uptake of the analogue and deoxythymidine varies from tissue to tissue (BAUGNET-MAHIEU and GOUTIER, 1968). The marked difference in phosphorylation rates of deoxythymidine and IUdR by various tissue extracts may account for the variations in incorporation. The iodopyrimidine is rapidly degraded by the tissues of mouse or man (WELCH and PRUSOFF, 1960) and a pathway for the degradation appears to involve cleavage of IUdR to iodouracil and subsequent dehalogenation with the formation of uracil and inorganic iodine (PRUSOFF, JAFFE and GUNTHER, 1960). Metabolism of IUdR and the metabolic sites for which there is evidence of its inhibition are indicated in Fig. 32.

There are presumably three major areas in which IUdR and its phosphorylated derivatives may exert inhibitory effects (PRUSOFF, 1967): (1) competition with substrates in several reactions involved in the incorporation of thymine into DNA; (2) negative feedback inhibition of dIUTP in reactions normally affected by dTTP;

and (3) incorporation of the analogue as a substitute for dTMP into cellular or viral DNA with a resultant effect on replication or expression of genetic information. DELAMORE and PRUSOFF (1962) investigated the effect of IUdR on the formation and utilization of phosphorylated derivatives of deoxythymidine in various murine and human neoplastic tissues. It was observed that the specific conversion primarily affected in the various tissues is a characteristic of the individual tissues. Thus, DNA

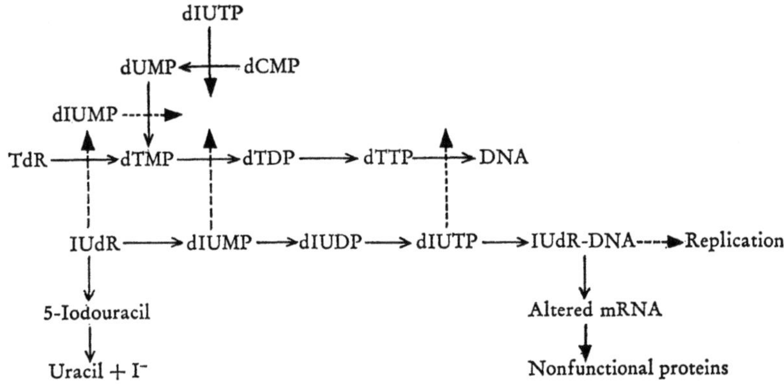

Fig. 32. Metabolism of IUdR and its potential sites of inhibition as indicated by ▶. Broken lines represent inhibitions which are not definitely established

polymerase is primarily inhibited in murine Ehrlich ascites carcinoma, human granulocytic leukemia and monocytic leukemia, whereas blockade of deoxythymidine kinase or dTMP kinase may be more important in murine L5178Y leukemia cells. Inhibition of these specific enzymes is probably mediated by the appropriate metabolite of the analogue, for example, inhibition of dTMP kinase by dIUMP or inhibition of DNA polymerase by dIUTP. It is of interest that while these enzymes are inhibited, the conversion of dTDP to dTTP is not inhibited in any of the tissues investigated.

PRUSOFF and CHANG (1968) studied the allosteric inhibition of dIUTP on deoxycytidylate deaminase derived from Ehrlich ascites carcinoma and L5178Y murine lymphoblastic cells. The inhibition produced is about 3—10 fold greater than that caused by dTTP suggesting a contributory role of dIUTP in suppression of cellular reproduction. However, it is realized that the significance of this inhibition depends on the pool of dIUTP formed as well as its depletion rate either by incorporation into DNA or by dephosphorylation by nucleoside triphosphatase (PRUSOFF, BAKHLE and SEKELY, 1965; PRUSOFF and CHANG, 1968).

Incorporation of IUdR into DNA has been shown to cause a significant increase in the temperature of helix-coil transition thereby indicating certain physiochemical changes in the structure of DNA. It is considered that possible impairment in the separation of the complementary DNA strands as a result of such structural changes may produce inhibition of DNA synthesis (CAMERMAN and TROTTER, 1964). Since replacement of deoxythymidine by IUdR in DNA of bacteria and phages leads to a rise in the incidence of mutation, a possibility also exists that incorporation of the analogue may increase the errors in base-pair formation during replication or transcription (GOZ and PRUSOFF, 1968). Such transcription errors find support in several reports which show that although viral DNA and viral antigens accumulate within the infected cells during propagation of viruses in the presence of IUdR, their assembly into viral particles does not occur (EGGERS and TAMM, 1966; SMITH and DUKES, 1964; KAPLAN, BEN-PORAT and KAMIYA, 1965). More specifically the results of KAPLAN and BEN-PORAT (1966) indicate that pseudorabies viral-DNA containing IUdR causes the synthesis of non-functional proteins which are incapable of virus assembly. Similarly, GOZ and PRUSOFF (1968) demonstrate that incorporation of IUdR into phage DNA in place of deoxythymidine results in decreased levels of activity of the phage-induced enzymes, dCMP hydroxymethylase and lysozyme.

The selective antiviral effect of IUdR has been investigated in several laboratories (HANNA and WILKINSON, 1965; SMITH, 1963; DEINHARDT, 1965; KAUFMAN, 1965). The results are contradictory and apparently fail to define the cause of selective activity. However, KAPLAN and BEN-PORAT (1967) observed that the selective antiviral property of IUdR is dependent on the concentration level of the analogue. It was shown that a considerable difference exists between the degree of incorporation of IUdR into the DNA of pseudorabies virus-infected and noninfected rabbit kidney cells when these cells are incubated with low concentrations (up to 1 mg/ml) of the analogue. These low concentrations of IUdR are capable of decreasing the viral infectivity by about 95 percent without any detectable inhibition of cell multiplication. This selectivity disappears when higher concentrations of IUdR are used.

Resistance: Strains of herpes simplex and vaccinia viruses which are resistant to IUdR have been shown to be deficient in the ability to induce the formation of deoxythymidine kinase. Deficiency of this enzyme results in limited conversion of the analogue to its phosphorylated form. Other mechanisms such as a decrease in sensitivity of the affected enzymes to inhibition or increase in their production have also been suggested from studies with a strain of resistant virus (PRUSOFF, 1967).

5-Bromo-2'-deoxyuridine (Bromodeoxyuridine)

5-Bromo-2'-deoxyuridine   Deoxythymidine

5-Bromo-2'-deoxyuridine (or BUdR), a structural analogue of deoxythymidine, has been shown to be a cytotoxic and mutagenic agent in a variety of biological systems (BROCKMAN and ANDERSON, 1963). The analogue has recently received special attention with regard to its therapeutic use as a radiosensitizer. The possible mechanisms for its increasing radiation sensitivity have been extensively investigated since the original report of this property in human cells by DJORDJEVIC and SZYBALSKI (1960).

Bromodeoxyuridine is phosphorylated to mono-(dBUMP), di-(dBUDP), and triphosphate (dBUTP) by kinases which are apparently indistinguishable from those which act on deoxythymidine and its phosphates, and is incorporated into DNA in place of deoxythymidine (BROCKMAN and ANDERSON, 1963; HITCHINGS and ELION, 1967).

The sequence of reactions in the incorporation of BUdR into DNA and the possible *loci* of inhibition are shown in Fig. 33. Since BUdR is anabolized like deoxythymidine to its nucleotide forms there is competition between natural and analogue substrates. Such competition may cause inhibition of thymine nucleotide synthesis to an extent dependent upon the concentrations of the analogue (BROCKMAN and

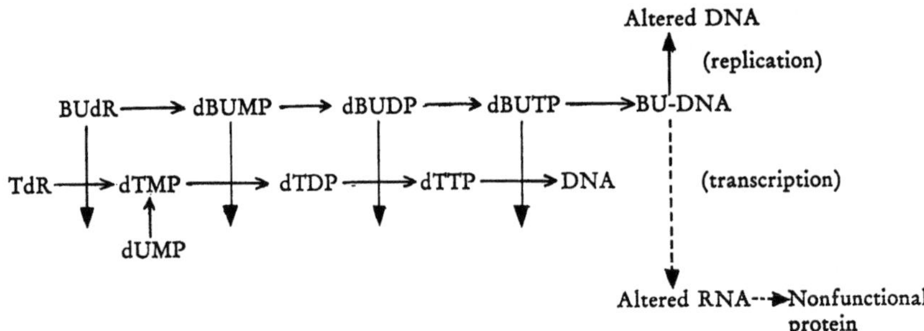

Fig. 33. Pathways of incorporation of BUdR into DNA. The possible *loci* of inhibition of these bromo derivatives at the nucleotide or incorporated level are shown by lines with ▶ Broken lines represent effects not definitely established

ANDERSON, 1963; HITCHINGS and ELION, 1967). However, the fact that in bacterial or mammalian cells BUdR replaces up to 100 percent of DNA deoxythymidine suggests that the biological effects may be predominantly associated with the changes in DNA. It is well documented that bromouracil incorporation into DNA results in increased mutation rate. The cause for this mutagenesis is believed to be faulty base-pairing between bromouracil and guanine instead of adenine during DNA replication, thus resulting in transition of bases in the new DNA chains. This point is supported by the observed aberrant guanine incorporation into polydeoxyribonucleotides in the *in vitro* DNA polymerase reaction with deoxyadenylic-deoxybromouridylic copolymer as template. The mutagenic effects of bromouracil incorporation have been discussed previously (BROCKMAN and ANDERSON, 1963). In the present discussion attention will be drawn to some recent works which emphasize the possibility of other physical alterations in DNA molecules due to bromouracil incorporation.

It is evident from recent findings that mutational effects are not solely responsible for the growth inhibition in the presence of BUdR. For example, GOULIAN, SINS-HEIMER and KORNBERG (1967) have shown that the minus strands of ΦX 174 virus can be produced in which all the thymine residues are replaced by bromouracil. These new strands are capable of producing active progeny viral DNA. The results do not support the transition of bases during this viral DNA replication but instead suggest that other mechanisms exist to explain the effects of bromouracil. GONTCHAROFF and MAZIA (1967) have studied the effect of introduction of BUdR into the DNA on the development of sea urchin embryos during early division stages. They observed that BUdR must be available at the time of DNA duplication to produce abnormalities. If BUdR is introduced into the DNA at any time up to the 8-cell stage, the embryos develop into one of two types of persistent abnormal blastulae but do not gastrulate and do not form echinochrome pigments. Incorporation of BUdR at a period between the 16-cell stage and about the 100 cell stage produces abnormal blastulae but does not affect echinochrome production. After about the 100 cell stage, the embryos are relatively insensitive to the analogue substitution. These results suggest that the modification of DNA in very early stages causes changes in developmental events whose phenotypic expression takes place much later. It is difficult to explain these effects by the mutagenic property of bromouracil. The authors suggest that these developmental abnormalities may relate more to the changes in the physical properties of DNA as a result of BUdR substitution rather than the primary nucleotide sequence. Similarly, ZAMECNIK and ZAMECNIK (1967) have shown that incorporation of BUdR into DNA sets off a chain of events which, in subsequent generations, results in persistent damage to steps involved in construction of chloroplasts in leaves of *Ageratum* plants. TOLIVER and SIMON (1967) showed that BUdR perturbs the cell cycle of bromouracil-tolerant HeLa cells by causing a nearly three-fold increase in the length of the S period, as compared to that of normal HeLa cells, but has little effect on the other parts of the cell cycle. Hypothetically this increased S period or slower rate of DNA synthesis may be based on the presence of different classes of thymine sites on the HeLa cell DNA molecule. During the inhibition of endogenous thymine synthesis, deoxythymidine and BUdR when present in approximately equal concentrations, compete more or less equally for sites on the DNA molecule (SIMON, 1963). But when the analogue is present in great excess, 25 percent of the sites are still occupied by thymine indicating that about 75 percent of the thymine sites will accept bromouracil as readily as thymine but that most of the remaining sites have preference for thymine. Thus, it is possible that in the presence of an excess BUdR, the synthesis of DNA would be delayed until a thymine moiety is inserted or a bromouracil moiety is forced in. Such forced substitution of bromouracil at certain thymine sites may alter the steric function of the DNA molecule. This "preferred site" hypothesis is supported by the experiments of HAUT and TAYLOR (1967) on BUdR substitution in DNA of bean roots *(Vicia faba)*. When bean roots are grown in solutions containing BUdR and an inhibitor of deoxythymidylate synthesis, DNA of hybrid density increases in amount during the first replication cycle and after some cells have entered the second replication cycle in BUdR, fully substituted DNA appears. However, in all preparations from roots grown in BUdR, a relatively large proportion of the analogue is incorporated into what appears to be partially substituted chains. When deoxythymidine is supplied after a period of growth in BUdR, the natural base is added almost exclusively to these partially substituted fractions of DNA indicating a strong preference for thymine in completion of the chains. Similarly, when limited substitution of BUdR occurs in the absence of an inhibitor of deoxythymidylate synthesis, the incorporation is almost exclusively into these partially substituted fractions and no hybrid with a fully substituted chain is produced under these conditions.

It is apparent from the above discussion that in addition to base-pairing errors there are certainly other physical or steric effects of BUdR incorporation into DNA. The molecular structure of BUdR, in fact, significantly differs from that of deoxythymidine (IBALL, MORGAN and WILSON, 1966). These physical effects probably contribute towards the marked increase in the sensitivity of the analogue-containing DNA to ultraviolet light, visible light, x-rays or radioactivity. The sensitization may be related to increased lability of the analogue-containing DNA to scission, its inability to undergo normal repair process or to photochemical changes at analogue and other sites in the polymer.

The effect of BUdR incorporation on protein synthesis has also been considered. Possible errors in transcription of bromouracil-containing DNA may result in inhibition or production of nonfunctional proteins. Although there is no direct experimental evidence for this, some reports have appeared which may support this view. For example, DUTTON, DUTTON and VAUGHAN (1960) observed a decrease in the rate of antibody synthesis by rabbit spleen cells which are exposed to BUdR. KIM et al. (1967) found that in HeLa cells exposed to BUdR the activity of DNA polymerase and uridine kinase is reduced by a large factor relative to changes in total cellular protein.

5-Hydroxyuridine

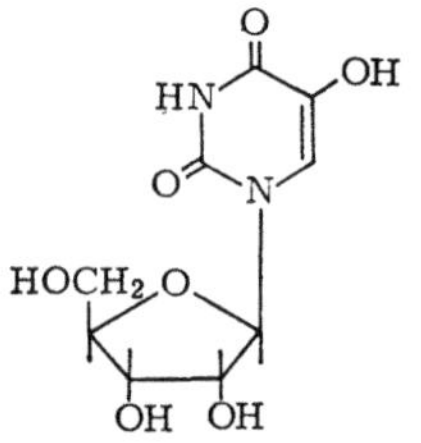

5-Hydroxyuridine 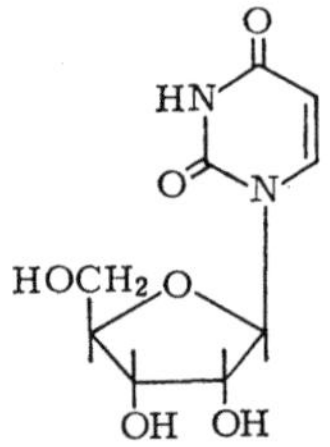   Uridine

The uridine derivative, 5-hydroxyuridine (or HO-UR) was first prepared by LEVENE and LAFORGE in 1912. Its synthesis was improved in the procedures of ROBERTS and VISSER (1952) and of UEDA (1960). The new method for the preparation of this nucleoside analogue also provides the mild conditions necessary for the preparation of its phosphate derivatives (VISSER, 1968; VISSER and ROY-BURMAN, 1968). The analogue is an effective inhibitor of bacterial, viral and tumor growths (SLOTNICK, VISSER and RITTENBERG, 1953; PEARSON, LAGERBORG and VISSER, 1956) and at low concentrations it has been reported to inhibit enzyme induction without affecting over-all protein synthesis in *Escherichia coli* (SPIEGELMAN, HALVORSON and BEN-ISHAI, 1955).

The metabolism of HO-UR has been studied in Ehrlich ascites tumor cells (SMITH and VISSER, 1965). Evidence has been obtained for the formation of its monophosphate (HO-UMP), diphosphate (HO-UDP), triphosphate (HO-UTP) and diphosphate sugar derivatives (HO-UDPX), and its incorporation into RNA. In studies to determine the quantitative differences in the metabolic fate of equimolar concentrations of uridine and HO-UR in Ehrlich ascites cells, it was found that the accumulated amounts in the respective acid-soluble nucleotides are not grossly different. However, the amount of uridine incorporated into RNA in these cells is 27 times greater than that of HO-UR. This inefficient incorporation of HO-UR into RNA is consistent with the subsequent finding that HO-UTP is a poor but specific substitute for UTP in the *E. coli* RNA polymerase reaction (ROY-BURMAN, ROY-BURMAN and VISSER, 1966). The low level of incorporation of HO-UMP into RNA has been related to low *pKa* of HO-UR, and explained, in part, by the failure of the polymerase to utilize the ionized form of the analogue. The analogue triphosphate is a substrate for UDP-glucose pyrophosphorylase obtained from yeast. The product formed, HO-UDP-glucose is oxidized by calf liver UDP-glucose dehydrogenase to yield HO-UDP-glucuronate (ROY-BURMAN, ROY-BURMAN and VISSER, 1968). The rate of oxidation, however, is much slower than that of UDP-glucose. In the presence of a crude enzyme preparation from yeast, HO-UTP reacts with xylose-1-phosphate to form HO-UDP-xylose (ROY-BURMAN, 1969). It is not known whether this UDP-xylose pyrophosphorylase activity is different from UDP-glucose pyrophosphorylase activity. The metabolism of HO-UR and the known inhibitory effects of its phosphorylated derivatives are summarized in Fig. 34.

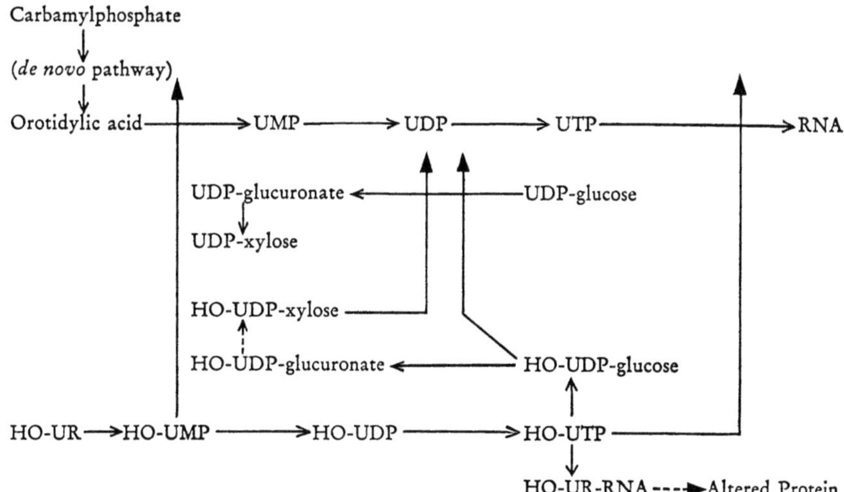

Fig. 34. Metabolism of HO-UR and the known inhibitory sites. Inhibition is indicated by ▶ and broken lines represent reactions not established

In *E. coli* 15T⁻, HO-UR inhibits both RNA and protein synthesis (BEN-ISHAI and VOLCANI, 1956). In Ehrlich ascites cells the effect of the analogue on RNA synthesis is much more pronounced than the effect on protein synthesis. The inhibitory effect on RNA synthesis has been explained in part by the demonstration that its phosphorylated derivative, HO-UMP strongly inhibits orotidylic acid decarboxylase of Ehrlich ascites cells (SMITH and VISSER, 1965). It was observed that the formation of uridine nucleotides from orotic acid is almost completely prevented by the presence of HO-UMP at a molar ratio of inhibitor to orotic acid of 5:1. Other reactions involved in the conversion of orotic acid to UTP are not affected by HO-UR or its phosphorylated products. The observation that there is a facile conversion of HO-UR to HO-UTP and some conversion to HO-UDP-sugars in the Ehrlich ascites cells, compared to a low level of incorporation into RNA, has led to studies of the effect of their derivatives on other potential enzyme reactions. These experiments have revealed additional sites of inhibition. The analogue triphosphate inhibits *in vitro* RNA synthesis catalyzed by *E. coli* RNA polymerase and acts as a competitive inhibitor of UTP in the polymerase reaction (ROY-BURMAN, ROY-BURMAN and VISSER, 1966). In contrast to the results on the incorporation of the analogue into RNA, the inhibitory effect is predominant when the pyrimidine moiety of HO-UTP exists in ionized form. Although HO-UMP strongly inhibits decarboxylation of orotidylic acid, the concentration of HO-UTP in whole Ehrlich ascites cells incubated in the presence of HO-UR is high compared to HO-UMP concentration. Thus, it is conceivable that inhibition of either the polymerase or decarboxylase reaction may predominate depending upon environmental conditions.

Some cofactor derivatives formed from HO-UTP may also contribute to growth inhibition. It has been shown that the oxidation of UDP-glucose by UDP-glucose dehydrogenase (reaction 11) is inhibited by HO-UDP-glucose (Roy-Burman, Roy-Burman and Visser, 1968). The cofactor analogue produces noncompetitive inhibition with respect to UDP-glucose and mixed type inhibition with respect to NAD+. At the enzyme optimum pH, $Ki/(Km$ for UDP-glucose) is about 200 while $Ki/(Km$ for NAD+) is 1.5. The values of these ratios indicate that the analogue is a more potent inhibitor of the binding of NAD+ to the enzyme than it is of UDP-glucose, thus accounting for the noncompetitive kinetics with respect to UDP-glucose. The mixed inhibition with respect to NAD+ is observed because the analogue not only interferes with binding of NAD+, but also competes with UDP-glucose since it is an alternate substrate for the reaction. Since UDP-glucose dehydrogenase of plant and animal origin is allosterically regulated by UDP-xylose (Neufeld and Hall, 1965), it was of interest also to investigate the effect of HO-UDP-xylose on this enzyme. A pathway for the formation of UDP-xylose follows the reaction sequence of UDP-glucose to UDP-glucuronate by dehydrogenation and UDP-glucuronate to UDP-xylose by decarboxylation. It has been shown that HO-UDP-xylose also inhibits UDP-glucose dehydrogenase, although to a lesser extent than UDP-xylose (Roy-Burman, 1969).

The demonstration that HO-UR is incorporated to a minor extent into RNA may account for the impairment of protein synthesis in some cases. It has been demonstrated that poly-5-hydroxyuridylic acid is an inefficient (Grunberg-Manago and Michelson, 1964) or completely ineffective (Roy-Burman, Roy-Burman and Visser, 1965) polynucleotide in its ability to direct phenylalanine polymerization. However, incorporation of amino acids other than phenylalanine into polypeptides may be stimulated by the polynucleotide analogue (Grunberg-Manago and Michelson, 1964). Thus, even though incorporation of HO-UR into RNA is not extensive, a small percentage of the analogue present in mRNA may have an inhibitory effect on protein synthesis or may result in production of altered proteins.

5-Aminouridine

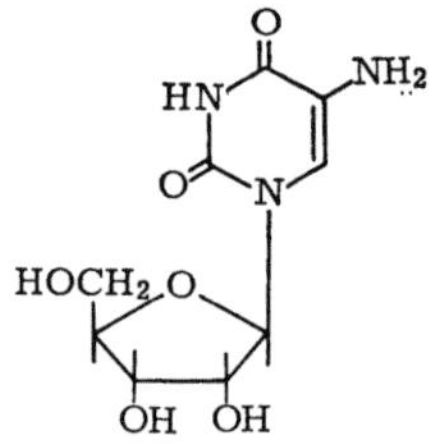

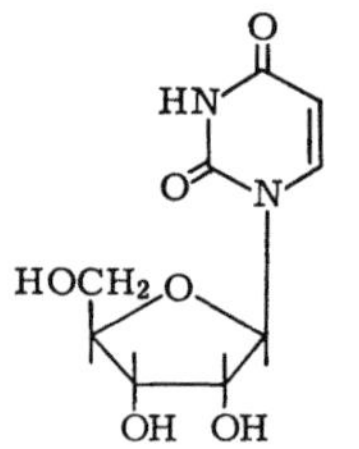

The pyrimidine nucleoside analogue, 5-aminouridine (or H₂N-UR) was prepared by Roberts and Visser (1952). The analogue, its free base and deoxyribonucleoside are known to inhibit growth of bacteria, fungi, viruses, protozoa and tumors (Kidder and Dewey, 1949; Roberts and Visser, 1952 a; Visser, Lagerborg and Pearson,

1952; BELTZ and VISSER, 1957; DUNN and SMITH, 1958; THEIL and ZAMENHOF, 1963). The base, 5-aminouracil has been shown to inhibit growth of plant roots (DUNCAN and WOODS, 1953; SMITH, FUSSELL and KUGELMAN, 1963) producing concomitant cytological alterations. It inhibits nuclear divisions while distorting chromosome structure and has been used to produce partial synchronization of nuclear division of *Vicia faba* root meristems (SMITH, FUSSELL and KUGELMAN, 1963).

In Ehrlich ascites cells H_2N-UR is phosphorylated to its monophosphate (H_2N-UMP), diphosphate (H_2N-UDP) and triphosphate (H_2N-UTP) derivatives and is incorporated into RNA and nucleoside diphosphate sugar compounds (SMITH, ROY-BURMAN and VISSER, 1966). The amount of H_2N-UR nucleotides formed is about half of uridine nucleotides formed under similar conditions. However, the amount of uridine incorporated into RNA is seven times greater than that of H_2N-UR. When supplied as 5-aminouracil or 5-aminodeoxyuridine the analogue is incorporated into bacterial DNA (WACKER et al., 1960; KABAT and VISSER, 1964).

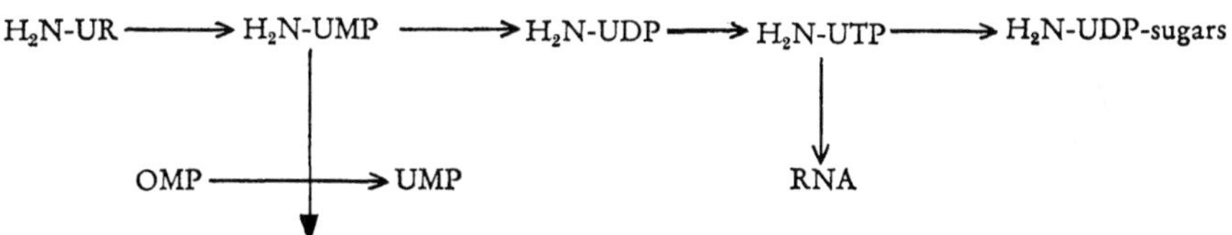

Fig. 35. Incorporation of H_2N-UR into RNA and cofactors and a known site of inhibition (indicated by ▶). Effects of its incorporation into RNA or cofactors are not known at present

Metabolism of H_2N-UR and its primary site of inhibition are shown in Fig. 35. Preliminary observations of the inhibition of carbamylphosphate incorporation into pyrimidines of nucleic acid and phosphate incorporation into phospholipids and nucleotides of rat liver slices or hepatoma by H_2N-UR indicated interference with the *de novo* pyrimidine nucleotide biosynthesis (WERKHEISER and VISSER, 1955; WERKHEISER, WINZLER and VISSER, 1955; EIDINOFF et al., 1959). Studies with Ehrlich ascites cell preparations showed that H_2N-UMP is a potent inhibitor of the conversion of orotic acid to uridine nucleotides accompanied by a large accumulation of orotidylic acid. Evidence has been obtained indicating that H_2N-UMP specifically inhibits orotidylic acid decarboxylase, the enzyme involved in the conversion of orotidylic acid to UMP (SMITH, ROY-BURMAN and VISSER, 1966). The results also suggested that the inhibition of this enzyme is probably of major importance at least early after H_2N-UR administration. Other factors may predominate after most of the inhibitor is phosphorylated beyond the monophosphate stage. For example, H_2N-UTP is a specific but inefficient substitute for UTP in reactions catalyzed by *E. coli* RNA polymerase and thus results in inhibition of RNA synthesis (ROY-BURMAN, ROY-BURMAN and VISSER, 1969). Incorporation of the analogue into nucleic acids may result in some functional alterations. It has been shown that Newcastle disease and Western equine encephalomyelitis viruses grown in the presence of H_2N-UR are hypersensitized to nitrous acid (THIRY, 1966). The seven-fold increase of this susceptibility to the deaminating agent may be explained presumably by a substitution of the analogue for uridine, which is not a target for nitrous acid. In its presence, the 6-methylpurine content of DNA from a thymine-requiring mutant of *E. coli* is increased as much as nine-fold (DUNN and SMITH, 1958; THEIL and ZAMENHOF, 1963). The base analysis of this DNA does not correspond to that expected from the Watson-Crick hypothesis (DUNN and SMITH, 1958). An explanation for these results is not apparent.

1-β-D-Arabinofuranosylcytosine (Arabinosylcytosine)

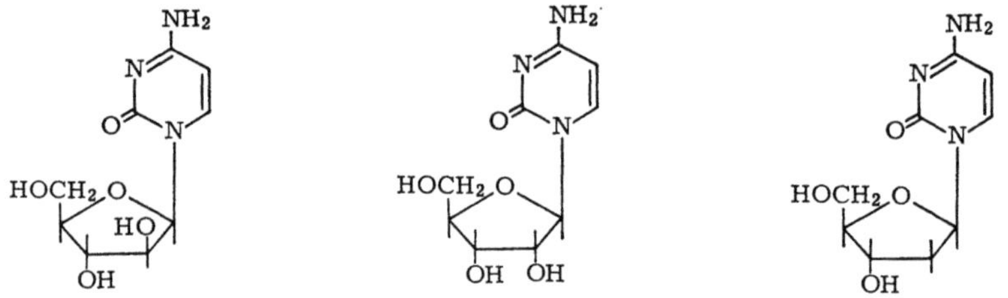

Arabinosylcytosine   Cytidine   Deoxycytidine

1-β-D-Arabinofuranosylcytosine (arabinosylcytosine or ara-C) is a synthetic nucleoside whose structure closely resembles the natural nucleosides, cytidine and deoxycytidine. The compound was first synthesized by WALWICK, ROBERTS and DEKKER in 1959. It inhibits the growth of tumor cells (EVANS et al., 1961; CHU and FISCHER, 1962; WODINSKY and KENSLER, 1965) and multiplication of DNA viruses in cell cultures (BUTHALA, 1964; LEVITT and BECKER, 1967). Although the analogue is effective in the control of herpes keratitis it seems to be slower than 5-iodo-2'-deoxy-uridine (IUdR) in clearing the initial infection (KAUFMAN, 1965; UNDERWOOD, ELLIOT and BUTHALA, 1965) and moreover, in contrast to IUdR, ara-C lacks the selective antiviral activity (KAUFMAN et al., 1964; BEN-PORAT, BROWN and KAPLAN, 1968). In clinical use ara-C has been effective in inducing partial remissions in a number of acute leukemia cases (CREASEY et al., 1966; HENDERSON and BURKE, 1965; HOWARD, CEVIK and MURPHY, 1966; TALLEY et al., 1967). In experimental immunological system it has delayed or prevented manifestations of the immune responses (KAPLAN et al., 1966; FISCHER, CASSIDY and WELCH, 1966). Pharmacological and enzymatic studies in man and mouse indicate that ara-C is rapidly deaminated to 1-β-D-arabino-furanosyluracil, an inactive metabolite (PAPAC et al., 1965; CAMIENER and SMITH, 1965; DOLLINGER et al., 1967).

It has been shown that deoxycytidine kinase is the enzyme which serves to phosphorylate ara-C to its monophosphate, ara-CMP in various animal and tumor tissues (CHU and FISCHER, 1965; SCHRECKER and URSHEL, 1968; GRINDEY, SASLAW and WARAVDEKAR, 1968; MOMPARLER and FISCHER, 1968; DURHAM and IVES, 1968). With the purified deoxycytidine kinase from calf thymus the Km (4.0×10^{-5} M) for ara-C is higher than the Km (1.4×10^{-5} M) for the natural substrate, deoxycytidine (MOMPARLER and FISCHER, 1968). It is of interest that ara-C is phosphorylated, in the presence of a dialyzed supernatant from the spleen of BDF_1 mice bearing advanced leukemia L1210 with UTP at twice the initial rate as compared to ATP at equimolar concentrations (GRINDEY, SASLAW and WARAVDEKAR, 1968). This indicates the presence of a deoxycytidine kinase in spleen with which UTP is the active phosphate donor and may relate to the observations that uridine, administered in conjunction with a suboptimal dose of ara-C, results in prolonged survival of BDF_1 mice bearing the leukemia in comparison with administration of the suboptimal dose of ara-C alone SASLAW et al., 1966; SASLAW et al., 1968). It is possible that the observed potentiating effect of uridine on the biological activity of ara-C is a result of increased phosphorylation of the analogue in the presence of UTP (GRINDEY, SASLAW and WARAVDEKAR, 1968). In biological systems, ara-CMP is further phosphorylated by kinases to ara-CDP and ara-CTP and the main intracellular metabolite of ara-C is the triphosphate (PIZER and COHEN, 1960; CARDEILHAC and COHEN, 1964; CHU and FISCHER, 1965; SCHRECKER and URSHEL, 1968). Ara-CTP does not act as a substrate in polynucleotide formation with DNA polymerase or RNA polymerase of animal cells or *E. coli* (CARDEILHAC and COHEN, 1964; FURTH and COHEN, 1967). Nor does *E. coli* polynucleotide phosphorylase incorporate ara-CDP into polynucleotides (CARDEILHAC and COHEN, 1964). However, it has been reported by some workers that radioactive ara-C is incorporated at very low levels into cellular RNA and DNA (SILAGI, 1965; CHU and FISCHER, 1965; CREASEY et al., 1966). Recently it has been reported that a purified DNA polymerase obtained from calf thymus can catalyze the incorporation of ^{3}H-ara-CTP into DNA to a minor extent (MOMPARLER, 1969). Most of the incorporation is at the 3′-hydroxyl terminal end and not within the polydeoxyribonucleotide chain.

Several studies have been carried out in various laboratories to elicit possible enzymatic sites of inhibition by ara-C and its phosphorylated derivatives. It has been demonstrated that administration of ara-C inhibits the synthesis of DNA in mammalian cells (CHU and FISCHER, 1962; KIM and EIDINOFF, 1964; LEVITT and BECKER, 1967; CHU and FISCHER, 1968) and in several strains of *E. coli* (PIZER and COHEN, 1960; SLECHTA, 1961). In connection with the studies of the effects of intracranial injection of ara-C on learning and retention of avoidance responding in gold fish, CASOLA et al. (1968) found that the analogue rapidly inhibited synthesis of DNA but not of RNA or protein. BEN-PORAT, BROWN and KAPLAN (1968) observed that ara-C inhibits the synthesis of DNA in noninfected rabbit kidney cells to a greater degree than in cells infected with herpes simplex virus or pseudorabies virus. Since virus multiplication is inhibited by the analogue to the same extent as DNA synthesis, they proposed that ara-C interferes with the infective process by inhibiting the DNA synthesis only. The inhibition of DNA synthesis in these uninfected cells can be overcome by deoxycytidine indicating prevention of effective inhibitor synthesis possibly by competition of deoxycytidine with ara-C at the level of phosphorylation (KAPLAN, BROWN and BEN-PORAT, 1968). The *in vitro* results of MOMPARLER and FISHER (1968), show that both deoxycytidine and ara-C are competitive inhibitors of each other for phosphorylation by deoxycytidine kinase, the former being much more inhibitory than the latter. The *Ki* value of ara-C in the inhibition of the phosphorylation of deoxycytidine is 3.6×10^{-5} M, whereas that of deoxycytidine in the inhibition of ara-C phosphorylation is 1.3×10^{-9} M. Recent findings of YOUNG and FISCHER (1968) of the action of ara-C on synchronously growing populations of mammalian cells reveal that the cytotoxicity caused by the analogue in S phase cells cannot be reversed by deoxycytidine, although this is reversed in cells during G1. The authors propose that ara-C acts specifically and irreversibly to kill cells during the S phase of the cell cycle and the unalterable lethality in S cells implies that the biochemical lesion caused by ara-C is closely involved with the process of DNA replication.

5*

Inhibition of DNA synthesis by ara-C may be explained by three possibilities, (1) phosphorylated derivatives of ara-C, particularly ara-CDP may interfere with the reductase system responsible for the conversion of pyrimidine ribonucleotides to deoxyribonucleotides thereby limiting the substrates for DNA synthesis; (2) the triphosphate derivative of ara-C may directly compete with dCTP in the DNA polymerase reaction; (3) small amounts of ara-CTP may be incorporated into DNA and such incorporation could be lethal for DNA replication. The first possibility has been considered by several workers to account for the growth inhibitions of animal neoplasms (CHU and FISCHER, 1962; EVANS et al., 1964; EVANS and MENGEL, 1964; KIM and EIDINOFF, 1965; KIMBALL et al., 1966). But it appears from the experiments of MOORE and COHEN (1967) that inhibition of ribonucleotide reduction does not play an important part in the inhibition of DNA synthesis by ara-C. They found that reduction of ribonucleotides by a partially purified enzyme system of rat tumor is only weakly inhibited by nucleotides of ara-C, to about the same degree as is observed with dCTP. Similarly KAPLAN, BROWN and BEN-PORAT (1968) reported that ara-C does not prevent reduction of CDP to dCDP in rabbit kidney cells incubated in medium free of deoxycytidine. However, when small amounts of deoxycytidine are supplied to these cells, the analogue appears to inhibit the reduction. The latter observation may be due to an increase in the intracellular pool of dCTP and an ensuing negative feedback effect.

The second possible mechanism of inhibition of DNA polymerase by ara-CTP has been investigated with the enzyme from *E. coli* (CARDEILHAC and COHEN, 1964) and from calf thymus (FURTH and COHEN, 1967; MOMPARLER, 1969). The bacterial polymerase is not inhibited by ara-CTP but the mammalian enzyme is inhibited. The inhibition appears to be competitive with dCTP, the Ki value being reasonably low (about 1×10^{-6} M). It should be mentioned that these studies also showed that neither bacterial nor the mammalian RNA polymerase are inhibited by ara-CTP consistent with the lack of inhibition of RNA synthesis *in vivo*. Thus, it appears that

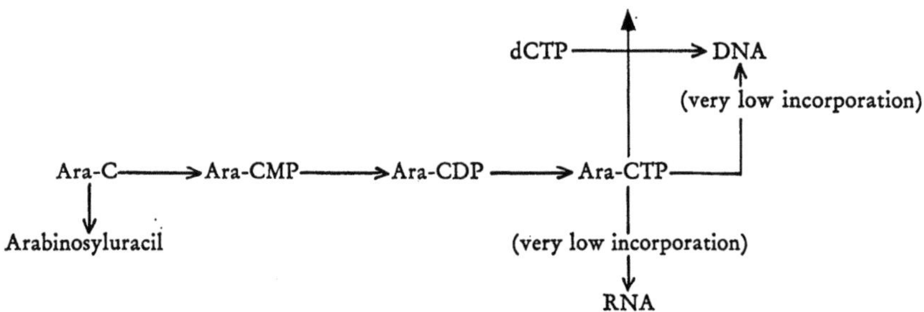

Fig. 36. Metabolism of Ara-C in animal cells and a known site of inhibition (indicated by ▶)

at least in animal cells inhibition of DNA polymerase may be the major mechanism of action. This is summarized in Fig. 36. In bacterial cells the mechanism of inhibition may be different. However it should be noted that ara-C does not inhibit *E. coli* at concentrations comparable to those at which it inhibits animal cells, i. e., $1—2 \times 10^{-6}$ M (FURTH and COHEN, 1968). It has been used at 4×10^{-3} M to inhibit DNA synthesis in *E. coli* (LARK and LARK, 1964).

The third hypothesis that the lethality of ara-C may be attributed to its incorporation into nucleic acids has been investigated by CHU and FISCHER (1965, 1968, 1968 a) and by COHEN and his workers (COHEN, 1966; FURTH and COHEN, 1967). CHU and FISCHER reported limited incorporation of the analogue into RNA and DNA fractions of murine leukemic cells and more recently provided evidence that the time-dependent logarithmic increment of the death of these cells is accompanied by the linear time-dependent incorporation of tritiated are-C into the acid soluble, DNA and RNA fractions. However, the amount of ara-C incorporated into nucleic acids is very low. Under the experimental conditions the estimated highest ratio of ara-C to cytidine or deoxycytidine is only about $1:2 \times 10^4$. COHEN and coworkers, on the other hand, have questioned the reliability of the detection of such low incorporation and also provided evidence contradicting the possibility of incorporation of ara-C into terminal positions which would block extension of polynucleotide chains. They showed that after extended inhibition of mouse fibroblasts by ara-C, mere separation of the inhibitor from the cells allows a rapid incorporation of thymidine into DNA. This indicates that inhibition of DNA synthesis by ara-C is probably not due to irreversible lesions in DNA resulting from terminal addition. Thus it may be concluded from the available data that incorporation of ara-C into nucleic acids is very low and at present the effect of such incorporation on cell multiplication is not understood at the biochemical level.

Resistance: CHU and FISCHER (1965) found that a clone of leukemic L5178Y cells selected for resistance to ara-C showed a considerable decrease in its capacity to phosphorylate the analogue or deoxycytidine. Similarly, the phosphorylation of ara-C and deoxycytidine is decreased over 97 percent in a resistant subline of ascitic leukemia L1210, although the capacity to phosphorylate cytidine or deoxythymidine remains unchanged (SCHRECKER and URSHEL, 1968). The fact that uptake of ara-C is not impaired in drug-resistant leukemia cells supports the view that the main criterion for ara-C sensitivity is the cellular ability for phosphorylation of the intracellular analogue (KESSEL, HALL and WODINSKY, 1967). With viral systems, BEN-PORAT, BROWN and KAPLAN (1968) reported that ara-C inhibits the synthesis of DNA in non-infected rabbit kidney cells to a greater degree than in cells infected with viruses of the herpes group and correlated this relative resistance with the decrease in the deoxycytidine kinase activity in the infected cells. Furthermore, it has been observed that the analogue does not induce a rise in the level of deoxycytidine kinase in the infected cells, whereas such an effect is produced in noninfected cells.

The mechanism of resistance appears to be different in an ara-C-resistant mutant of L5178Y murine leukemia recently isolated by MOMPARLER, CHU and FISCHER (1968). This resistant mutant, when compared to parent cells, maintains almost the same capacity to phosphorylate ara-C in both whole cell or cell-free systems. Any increase in deamination or cleavage of the analogue is not aparent in the resistant cells. However, formation of an increased pool size of deoxycytidine nucleotides has been observed in the resistant cells. It is possible that the mutant cells have become resistant to ara-C by avoiding the inhibitory effect of ara-CTP through the production of excess dCTP and effectively competing with ara-CTP formation.

Chapter 4

Nucleoside Antibiotics

The antibiotics which resemble nucleosides in their structures have been termed nucleoside antibiotics. These natural products can be separated into two major groups, one containing amino acid-linked compounds such as puromycin, lysylamino-adenosine, homocitrullylamino-adenosine, blasticidin S etc., and the other consisting of adenosine-like compounds which do not contain amino acids. The antibiotics of the former group, in general, interfere with protein synthesis in biological systems and do not function as structural analogues of nucleosides. The members of the latter group with a few exceptions such as psicofuranine and decoyinine are converted to nucleotide derivatives in cells and participate in many reactions of adenosine and its nucleotides. These adenosine-like antibiotics may be considered as both structural and functional analogues of nucleic acid components and thus warrant inclusion in the present subject of discussion.

Interest in the potential chemotherapeutic activity of the nucleoside antibiotics has led to efforts toward an understanding of their mechanism of action. This chapter emphasizes the importance of these investigations and provides the background for future progress in this direction. A new nucleoside antibiotic, namely showdomycin, which is structurally related to uridine, has also been included for its interesting and different mechanism of action.

The chemistry and biochemistry of the amino acid-containing or adenosine-like nucleoside antibiotics have been considered by Fox et al. (1966). A recent volume edited by GOTTLIEB and SHAW (1967) considers their mechanism of action. These articles should be consulted for further details.

Tubercidin

Tubercidin   Adenosine

The antibiotic tubercidin was isolated from culture filtrates of *Streptomyces tubercidicus* (ANZAI, NAKAMURA and SUZUKI, 1957; NAKAMURA, 1961). Its chemical structure has been established as 7-deazaadenosine or 4-amino-7β-D-ribofuranosyl-pyrrolo-

(2,3-d) pyrimidine (SUZUKI and MARUMO, 1960, 1961). The original observation of th growth inhibitory effects of tubercidin on *Mycobacterium tuberculosis* and on *Candida albicans* was followed by a series of reports about its toxic effects on various mammalian cell strains grown *in vitro* and its activity against some experimental tumors (SMITH, LUMMIS and GRADY, 1959; BRINDLE et al., 1961; BHUYAN, RENIS and SMITH, 1962; RENIS, JOHNSON and BHUYAN, 1962; DUVALL, 1963; BLOCH et al., 1964; OWEN and SOUTH, 1964; ACS, REICH and MORI, 1964). Recently BLOCH, LEONARD and NICHOL (1967) reported that the growth of *Streptococcus faecalis* is markedly inhibited by tubercidin. A fifty percent inhibition is produced at 2×10^{-8} M concentration of the inhibitor.

The glycosidic bond of tubercidin, unlike that of adenosine is resistant to acid hydrolysis. This property of the antibiotic is consistent with its resistance to cleavage by adenosine posphorylase (BLOCH, LEONARD and NICHOL, 1967). Tubercidin is a substrate for adenosine kinase and is converted to tubercidin 5′-monophosphate (TuMP). The reported Km value for tubercidin with partially purified adenosine kinase of rabbit liver is 4.0×10^{-5} M, whereas that for adenosine is 1.6×10^{-6} M (LINDBERG, KLENOW and HANSEN, 1967). Because of the inertness of tubercidin towards degradation by adenosine phosphorylase or deamination by adenosine deaminase, the conversion to TuMP is quite efficient. Further phosphorylations of TuMP to its di-(TuDP) and triphosphate (TuTP) are carried out by appropriate kinases. In cell-free preparations of Ehrlich ascites cells TuMP is almost completely converted to TuTP (ROY-BURMAN, unpublished results). The antibiotic is incorporated in limited amounts into DNA and RNA of mouse fibroblasts and of *S. faecalis* (ACS, REICH and MORI, 1964; BLOCH, LEONARD and NICHOL, 1967). NISHIMURA, HARADA and IKEHARA (1966), however, observed that in RNA synthesis *in vitro* by *E. coli* RNA polymerase, little TuTP is utilized in the presence of natural DNA or some polydeoxyribonucleotides with ordered nucleotide sequence. Incorporation of tubercidin into DNA indicates formation of deoxyribonucleotides of tubercidin. In fact it has been shown recently that TuDP and TuTP are substrates for ribonucleotide reductase systems from *E. coli* and *L. leichmanni*, respectively (CHASSY and SUHADOLNIK, 1968; SUHADOLNIK, FINKEL and CHASSY, 1968).

Due to its close structural resemblance to adenosine, tubercidin and its derivatives can participate in many other reactions of adenine nucleotides. Thus it appears that substitution of the N-7 of adenosine by a methylene group does not change the substrate activity in a wide variety of enzyme reactions. Incorporation of tubercidin into nicotinamide-deazaadenine dinucleotide, analogous to the formation of NAD, has been reported (BLOCH, LEONARD and NICHOL, 1967). The triphosphate derivative is an effective substrate for rabbit liver tRNA with adenosine nucleoside terminus pyrophosphorylase by which tubercidin is incorporated at the 3′-termini of tRNA molecules. Amino acid acceptor and transfer activities of these analogue-containing tRNA molecules are not much different from their normal counterparts (URETSKY et al., 1968). Polynucleotide phosphorylases utilize TuDP for *in vitro* formation of homo- or heteropolymers of TuMP. The homopolymer (poly-Tu), like polyadenylic acid, functions as a template for polylysine synthesis (URETSKY et al., 1968). Similarly, the synthetic trinucleotides, TupApA and TupCpC stimulate the binding of ribosomes, as expected, to lysyl- and threonyl-tRNA respectively (IKEHARA and OHTSUKA, 1965). Other biochemical reactions in which appropriate derivatives of tubercidin act as substrates are those catalyzed by yeast hexokinase, rabbit muscle myokinase, phosphoenolpyruvate kinase, myosine ATP-ase and adenosine kinase (as phosphate donor) (Acs and REICH, 1967).

It is noted earlier that adenosine phosphorylase and adenosine deaminase are ineffective on tubercidin. It is also known that TuTP can neither inhibit or replace ATP as the energy source for amino acid activation. This may suggest that the binding of ATP to amino acid activating enzymes is mediated in part by an interaction involving N-7 of adenine (URETSKY et al., 1968).

The sequence of reactions leading to the incorporation of tubercidin into nucleic acids and cofactors and the possible effects of such incorporations are outlined in Fig. 37. The lethal effect of the antibiotic on mammalian cells may be explained in

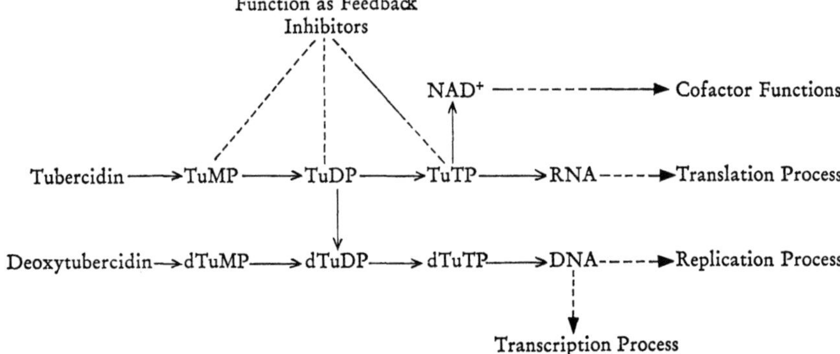

Fig. 37. Sequence of reactions in the incorporation of tubercidin into nucleic acids and cofactors. Broken lines with ▶ indicate the possible *loci* of interference with normal metabolic processes

part by its incorporation into nucleic acids. Acs and REICH (1967) observed that when mouse fibroblast cells are exposed to tubercidin certain morphological changes occur which resemble viral cytogenic effects. These tubercidin-induced changes are prevented by actinomycin. Since actinomycin does not inhibit the cellular uptake of tubercidin, its phosphorylation or incorporation into DNA, it is likely that actinomycin effects are associated with blocking of tubercidin incorporation into RNA. The growth of mengovirus in L-cells is inhibited by tubercidin and it has been shown that the antibiotic is incorporated into this viral RNA (Acs, REICH and MORI, 1964). Although the results indicate a direct relationship between inhibition and incorporation RNA, these cannot explain all the cellular effects of tubercidin. For example, rapid and almost simultaneous inhibition of the cellular synthesis of DNA, RNA and protein by the antibiotic cannot be due only to the fraudulent RNA production. Incorporation into DNA is a possibility which has not been studied in detail. However, the lethal effect on DNA is illustrated by the incorporation of the antibiotic, when supplied as deoxytubercidin into DNA of mouse fibroblasts followed by irreversible loss of viability of these cells (Acs, REICH and MORI, 1964).

BLOCH, LEONARD and NICHOL (1967) have suggested that the impairment of the growth of *S. faecalis* by tubercidin is mainly due to its interference with the utilization of glucose in this organism. This suggestion is based on the observation that the inhibitory effect of the antibiotic is prevented by purine and pyrimidine nucleosides, ribose-5-phosphate, pyruvate, and certain amino acids. Furthermore, in the presence of the antibiotic the organism uses pyruvate instead of glucose, whereas in its absence, glucose serves as the main source of energy. The actual sites of inhibition in the glycolytic pathway are not known. A possible mechanism which has been considered is interference with pyridine nucleotide dependent reactions because of the known facile conversion of tubercidin to TuTP and incorporation of TuTP into nicotinamide-deazaadenine dinucleotide.

Several other *loci* of action may be conceived, although there is no experimental evidence for these at present. The close resemblance between tubercidin and adenosine would permit competitive retardation of many reactions in the synthesis or utilization of adenine nucleotides by the appropriate tubercidin nucleotides. The cumulative effect may contribute significantly to the inhibition of cellular processes. Inhibition of multiple cellular functions by the accumulated tubercidin nucleotides through repression or allosteric mechanisms can also be speculated. Although tubercidin corresponds to adenosine in Watson-Crick type base-pairing, it is deficient in further hydrogen bond forming ability. This has been brought out by the studies of IKEHARA and FUKUI (1968). They showed that poly-Tu has a broad melting profile like poly-A and with poly-U it can form a 1:1 complex like poly-A:poly-U. However, poly-Tu is incapable forming a 1:2 complex with poly-U. This is consistent with the lack of N in the structure of tubercidin at position N(7) corresponding to adenine. It is believed that in poly-A: 2 poly-U a hydrogen bond to the second poly-U chain involves N(7) of adenine. The occurrence or role of triple stranded polynucleotide structures in biological systems is not known. In the event these do exist, incorporation of tubercidin into these structures may have significant effects.

In summary, it can be said that tubercidin, being a close structural analogue of adenosine, participates in many biological reactions. Its adverse effects on growth are related to its incorporation into nucleic acids and cofactors, and to a mechanism by which it can produce rapid and profound inhibition of protein synthesis. Further investigations are required for a clear understanding of these effects at the molecular level.

Toyocamycin

Toyocamycin

Adenosine

Toyocamycin was isolated from *Streptomyces toyocaensis* by NISHIMURA et al. (1956). Its chemical structure is 4-amino-5-cyano-7β-D-ribofuranosyl-pyrrolo-(2,3-d) pyrimidine (OHKUMA, 1961). Thus toyocamycin is structurally very similar to tubercidin, differing only in the presence of a nitrile group at 5 position. This antibiotic inhibitis *Candida albicans* and *Mycobacterium tuberculosis*, without notable action on other microorganisms such as many gram-positive and gram-negative bacteria, fungi and yeast. It has exhibited inhibitory effects on the growth of some animal tumors (SANEYOSHI, TOKUZEN and FUKUOKA, 1965).

Toyocamycin acts as a substrate for mammalian adenosine kinase. The reported Km is 8.0×10^{-6} M whereas that for adenosine is 1.6×10^{-6} M (LINDBERG, KLENOW and HANSEN, 1967). It is phosphorylated by Ehrlich ascites tumor cells to the mono-(ToMP), di-(ToDP) and triphosphate (ToTP) derivatives, and is incorporated in limited amounts into RNA of these cells (SUHADOLNIK, UEMATSU and UEMATSU, 1967). Small amounts of incorporation of the antibiotic into DNA of these cells indicate the formation of deoxyribonucleotides of toyocamycin. This is supported by the recent findings that ToDP and ToTP are substrates for ribonucleotide reductase systems from *E. coli* and *L. leichmanni* respectively (CHASSY and SUHADOLNIK, 1968; SUHADOLNIK, FINKEL and CHASSY, 1968).

The mechanism of action of toyocamycin is not known at present although some speculations can be made. The antibiotic is closely related to tubercidin in structure and since it is metabolized like adenosine including incorporation into nucleic acids, it appears likely that the inhibitory effects may be analogous to those produced by tubercidin. Toyocamycin, however contains a cyano group at C-5 and this substitution may impose steric as well as functional factors in its participation at certain enzymatic levels. It is known, for example, that ToTP is about thirty percent as efficient as ATP as a phosphate donor with ATP: phosphoglycerate transphosphorylase but is not a phosphate donor with luciferase (SUHADOLNIK, UEMATSU and UEMATSU, 1967).

Sangivamycin

Sangivamycin was discovered by RAO and RENN (1963) in culture filtrates of a species of *Streptomyces*. Its structure was reported by RAO (1965) to be 4-amino-5-carboxamido-7β-D-ribofuranosyl-pyrrolo-(2,3-d) pyrimidine. Sangivamycin is the third antibiotic of the pyrrolopyrimidine ribonucleoside series and it bears a close structural resemblance to tubercidin and toyocamycin. Antitumor activity of sangivamycin has been demonstrated in HeLa cell cultures and in mouse leukemia L1210 (RAO and RENN, 1963).

The antibiotic is readily converted by mammalian cells to its nucleotides (ACS and REICH, 1967). Sangivamycin triphosphate acts as a specific substitute for ATP in reactions catalyzed by RNA polymerase of *Micrococcus lysodeikticus*, and is incorporated into RNA (SUHADOLNIK et al., 1968). *In vitro* studies on the formation of sangivamycin deoxyribonucleotides have shown that although sangivamycin triphosphate functions as a substrate for the ribonucleotide reductase system from *L. leich-*

manni, the diphosphate derivative is inactive with the enzyme system from *E. coli* (SUHADOLNIK, FINKEL, and CHASSY, 1968; CHASSY and SUHADOLNIK, 1968).

No information concerning the mode of action of sangivamycin is presently available. Its pyrrolopyrimidine structure and behavior as an adenosine analogue may indicate some similarities in the activities of sangivamycin, toyocamycin and tubercidin. Further investigations on the action of these three pyrrolopyrimidine antibiotics would be of considerable interest to correlate the structural changes at C-5 from —H to —CN to —$CONH_2$ with the alterations in functional abilities.

Formycin

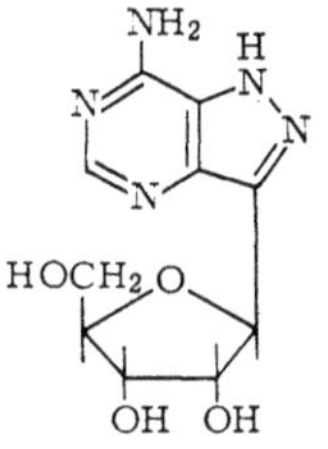

Formycin 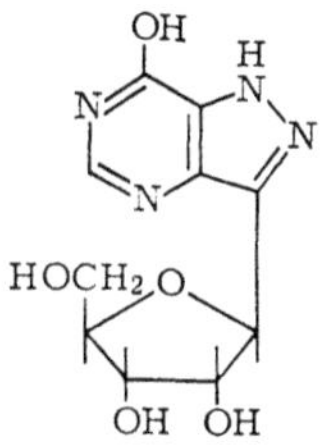   Formycin B 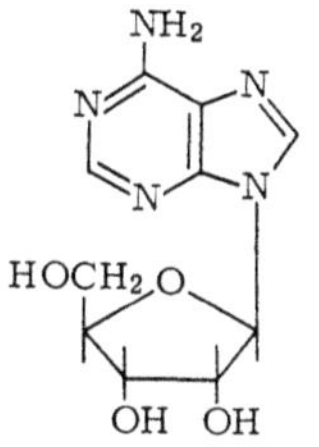   Adenosine

Formycin (or formycin A) was isolated by HORI et al. (1964) from culture filtrates of the actinomycete, *Nocardia interforma.* The structure of this antibiotic is 7-amino-3β-D-ribofuranosyl-pyrazolo-(4,3-d) pyrimidine (KOYOMA et al., 1966; ROBINS et al., 1966. Formycin has hown growth-inhibitory effects on various neoplastic cells, *Xanthomonas oryzae* and viruses (HORI et al., 1964; TAKEUCHI et al., 1966). The inhibitory activity of the antibiotic is more pronounced on solid forms of Ehrlich carcinoma and L-1210 than on ascites forms. Among the various organs of mice, formycin is predominantly concentrated in spleen and kidney after subcutaneous injection (ISHIZUKA et al., 1968).

In Ehrlich ascites cells formycin is phosphorylated to its mono- (FoMP), di- (FoDP) and triphosphate (FoTP) derivatives (CALDWELL, HENDERSON and PATERSON, 1966; UMEZAWA et al., 1967). The triphosphate, FoTP acts as a substrate for ATP in the *E. coli* RNA polymerase reaction, although the rate of RNA synthesis is somewhat slower in its presence (IKEHARA et al., 1968). It should be noted, however, that phosphorylations of formycin in *E. coli* is very low. UMEZAWA et al. (1967) have suggested that the rate of deamination of the antibiotic relative to the rate of its phosphorylation may be higher in *E. coli* than in tumor cells. The product of deamination of formycin is formycin B which was also isolated from *N. interforma* by KOYOMA and UMEZAWA (1965). Formycin B has been identified as 7-hydroxy-3β-D-ribofuranosyl-pyrazolo-(4,3-d) pyrimidine (KOYOMA et al., 1966). Formycin B is almost ineffective in inhibiting tumor growths but possesses significant activity against *Xanthomonas oryzae* and some viruses (KOYOMA and UMEZAWA, 1965; TAKEUCHI et al., 1966). Another metabolite, oxyformycin B, appears in urinary excretions of rabbits injected with formycin A or B. Oxyformycin B is biologically inert and is probably an end product of the detoxication mechanism (ISHIZUKA et al., 1968 a). Formation of this product from formycin B is catalyzed by aldehyde oxidase (TSUKADA et al., 1969).

The present state of knowledge is inadequate to outline the mechanism of action of formycin. There is evidence, however, which indicates that the inhibition produced by the antibiotic is dependent on its phosphorylations. Synthesis of PRPP in Ehrlich ascites cells is inhibited by formycin and presumably the active derivative is FoTP which interferes with the utilization of ATP in this reaction (HENDERSON et al., 1967). It is also possible that many other reactions requiring participation of ATP may be affected by FoTP. As far as base-pairing capacity is concerned, it appears that formycin behaves like adenosine. IKEHARA et al. (1969) studied the binding of amino-acyl transfer RNA to ribosomes using polyribonucleotides containing formycin. They observed that the binding of histidyl-tRNA is stimulated by the copolymer poly-(Fo-C) with about half the efficiency of poly-(A-C), while threonyl-tRNA is stimulated equally by poly-(Fo-C) and poly-(A-C). In *E. coli* protein synthesizing systems, poly-(Fo-C) and poly-(Fo-G) direct synthesis of poly-(his-thr) and poly-(arg-glu), respectively with no observed mistranslations. The authors suggest that incorporation of formycin into mRNA may not cause miscoding at translation levels. While this may be true in the case of heteropolymers containing the antibiotic, poly-Fo itself fails to code for the synthesis of polypeptides (WARD and REICH, 1968). An explanation for this inactivity may be obtained from the following discussion.

Physical and enzymatic studies of formycin containing polymers have suggested that formycin residues may exist in either a *syn* or *anti* conformation (WARD and REICH, 1968). The structure of formycin as shown on page 75 is the *syn* conformation. The *anti* conformation is achieved by rotating the chromophore 180° on the glycosyl bond. The tendency of formycin to adopt the *syn* conformation in poly-Fo accounts for the anomalies of the structure and function of this polynucleotide. Since the *anti* conformation is indispensible for ordinary base-pairing, the formycin residues are *anti* in ordered structures of the Watson-Crick type. It is a mixture of *syn* and *anti* in single-stranded polymers and entirely *syn* in its homopolymer (WARD and REICH, 1968).

Since formycin is partly deaminated to formycin B it is equally important to consider the biological effects of the latter. Formycin B is reported to inhibit nucleoside metabolism in *Xanthomonas oryzae* by interfering with some processes essential for entry of exogenous nucleoside into these cells (HORI et al., 1968). SHEEN, KIM and PARKS (1968) reported that the antibiotic is a competitive inhibitor of purified erythro-cytic purine nucleoside phosphorylase. The observed values of Km and Ki are 5×10^{-5} M and 1×10^{-4} M respectively. Evidence is also obtained indicating that formycin B inhibits this enzyme both in intact human erythrocytes and in intact murine sarcoma 180 cells. Since formycin B is neither converted to its phosphorylated derivatives nor degraded by nucleoside phosphorylase (because of its carbon-carbon nucleosidic linkage), it appears that the inhibition of purine nucleoside phosphorylase could be extensive in some biological systems. But this effect of formycin B cannot be general. For example, when the antibiotic is administered directly it does not cause inhibition of Ehrlich carcinoma or mouse leukemia L-1210 while formycin inhibits their growth (ISHIZUKA et al., 1968).

Resistance: A subline of Ehrlich ascites carcinoma has been investigated by CALD-WELL, HENDERSON and PATERSON (1966 a) for its resistance against formycin. Cells of this particular subline were originally selected for resistance to 6-mercaptopurine ribonucleoside; these cells are known to be deficient in purine nucleoside kinase activity. The parent line of these cells which is sensitive to inhibition by formycin converts the nucleoside to its mono-, di- and triphosphate derivatives of which triphosphate is the most abundant of the three nucleotides. The cells of the resistant subline show inefficiency in these conversions as only traces of these nucleotides are detected in cell extracts. Thus the loss of cellular ability to convert the antibiotic to its active phosphorylated form appears to be a mechanism of resistance formation against formycin.

Cordycepin

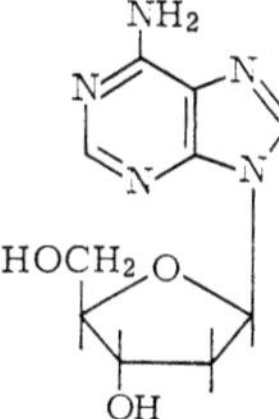

Cordycepin
(3'-Deoxyadenosine)   Adenosine   2'-Deoxyadenosine

Cordycepin was isolated as a metabolic product of the mold, *Cordyceps militaris* (Linn.) by CUNNINGHAM et al. (1951). The heterocyclic part of the compound was identified as adenine by BENTLEY, CUNNINGHAM and SPRING in 1951 and recently the complete structure was shown to be 3'-deoxyadenosine (KACZKA et al., 1964, 1964 a; SUHADOLNIK and CORY, 1964; FREDERIKSEN, MALLING and KLENOW, 1965). Cordycepin inhibits the growth of bacterial and mammalian cells both *in vitro* and *in vivo* (CUNNINGHAM et al., 1951; JAGGER, KREDICH and GUARINO, 1961; ROTTMAN and GUARINO, 1964; KACZKA et al., 1964).

The present state of knowledge on the mode of action of cordycepin has been reviewed (GUARINO, 1967). The antibiotic acts as a substrate for partially purified mammalian adenosine kinase. The reported Km for cordycepin is 4.6×10^{-4} M as compared to 1.6×10^{-6} M for adenosine (LINDBERG, KLENOW and HANSEN, 1967). In suspensions of Ehrlich ascites cells it is converted to mono-, di- and triphosphate derivatives (KLENOW, 1963). SHIGEURA and BOXER (1964) reported a limited incorporation of the antibiotic from its triphosphate (CoTP) into terminal positions of RNA chains formed in the presence of *Micrococcus lysodeikticus* RNA polymerase. Small amounts of incorporation into RNA and DNA of human tumor cells in culture have been observed (CORY et al., 1965). The inability of the ribonucleotide reductase system of *L. leichmanni* to reduce CoTP to the corresponding 2'-deoxy form (SUHADOLNIK, FINKEL and CHASSY, 1968) may indicate direct incorporation of CoTP into DNA. In the degradative reactions cordycepin is broken down to adenine and 3-deoxyribose and deaminated to 3'-deoxyinosine (GUARINO, 1967). The reported Km for cordycepin with mammalian adenosine deaminase is 2.2×10^{-5} M whereas that for adenosine is 3.5×10^{-5} M (CODDINGTON, 1965).

The metabolic conversions of cordycepin and the possible *loci* for inhibitory action are outlined in Fig. 38. The phosphorylated forms of the antibiotic are involved in the inhibition of several reactions. The monophosphate, CoMP is an inhibitor of phosphoribosyl-pyrophosphate amidotransferase isolated from *B. subtilis* or pigeon liver (ROTTMAN and GUARINO, 1964 a). This enzyme catalyzes an early step of purine biosynthesis, namely, formation of 5-phosphoribosylamine from PRPP. It is normally regulated through inhibition by AMP. OVERGAARD-HANSEN (1964) obtained evidence to indicate inhibition by CoTP of the synthesis of PRPP from ribose-5-phosphate

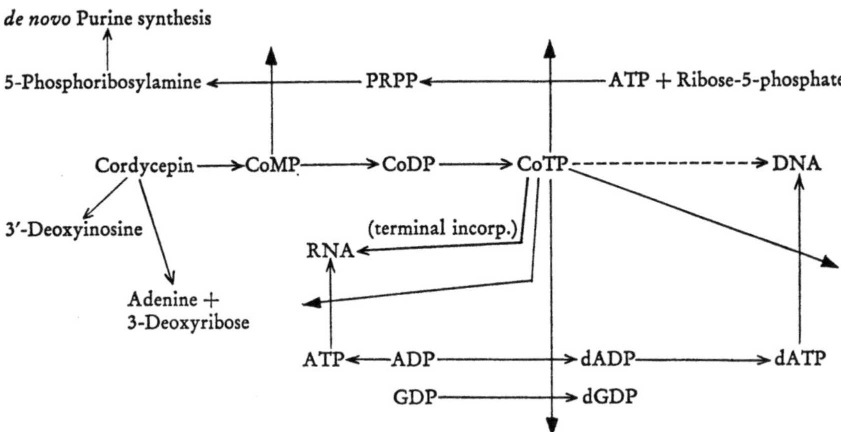

Fig. 38. Metabolic conversions and the possible inhibitory sites (indicated by ▶) of cordycepin

and ATP in extracts of Ehrlich ascites tumor cells. The triphosphate also interacts at the allosteric site of ribonucleotide reductase of *E. coli* and modifies the enzyme such that the reductions of ADP and GDP are both inhibited. The observed inhibitions are about one-half of those produced by dATP which is a normal regulator of these reductions (CHASSY and SUHADOLNIK, 1968). Similarly, with the reductase system from *L. leichmanni* CoTP mimics dATP in inhibiting reductions of ATP and CTP (SUHADOLNIK, FINKEL and CHASSY, 1968).

Synthesis of RNA in Ehrlich ascites cells is strongly inhibited by CoTP (KLENOW and FREDERIKSEN, 1964; FREDERIKSEN and KLENOW, 1964). Inhibition of RNA polymerase of *M. lysodeikticus* has been observed with CoTP while CoDP and CoMP have no effect (SHIGEURA and GORDON, 1965). Synthesis of DNA by *E. coli* DNA polymerase is also inhibited by the antibiotic triphosphate (SUHADOLNIK and CORY, 1965), although the growth of *E. coli* is not inhibited by this antibiotic. Inhibition of polymerases may also result from incorporation of CoMP into 3'-OH ends of growing polynucleotide chains. Such incorporation may not support further elongation of chains because of the absence of the 3'-OH group. This hypothesis is supported by the observation that additions of increasing amounts of CoTP in the absence of ATP in *M. lysodeikticus* RNA polymerase reaction mixtures do not significantly stimulate RNA synthesis (SHIGEURA et al., 1966). Similarly, with a RNA polymerase preparation of Ehrlich ascites cells, TRUMAN and KLENOW (1968) observed that an initial stimulation of RNA production by CoTP is followed by a strong inhibition. Internal incorporation of the antibiotic into nucleic acids has not been definitely established. In case internal incorporation occurs even in small amounts, some unique effects on structural organizations of nucleic acids may be expected since cordycepin must be bound in the polymers through a 2'—5' rather than the normal 3'—5' internucleotide linkage.

3'-Amino-3'-deoxyadenosine

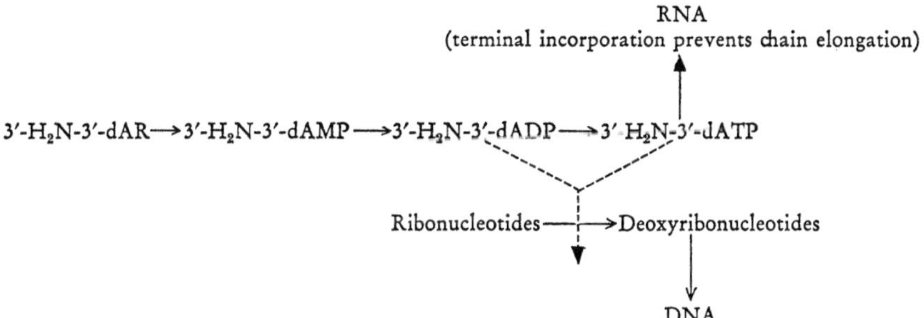

3'-Amino-3'-deoxyadenosine (3'-H_2N-3'-dAR) was isolated from culture filtrates of the turf fungus *Helminthosporium* sp. no. 215 by GERBER and LECHEVALIER (1962). Prior to this report of its natural occurrence the compound was synthesized by BAKER. SCHAUB and KISSMAN (1955) in connection with their studies on puromycin. The antibiotic has shown antitumor activity against a variety of ascites tumors in mice (PUGH, LECHAVALIER and SOLOTOROVSKY, 1962; PUGH and GERBER, 1963).

In Ehrlich ascites cells 3'-H_2N-3'-dAR is phosphorylated to its mono-, di- and triphosphate derivatives (SHIGEURA et al., 1966; TRUMAN and KLENOW, 1968). It acts as a substrate for partially purified mammalian adenosine kinase, the reported *Km* being 6.1×10^{-4} M, whereas that for adenosine is 1.6×10^{-6} M (LINDBERG, KLENOW and HANSEN, 1967).

Phosphorylations of 3'-H_2N-3'-dAR and the possible inhibitory reactions are outlined in Fig. 39. These results are obtained from studies with Ehrlich ascites cells. The

RNA
(terminal incorporation prevents chain elongation)

3'-H_2N-3'-dAR→3'-H_2N-3'-dAMP →3'-H_2N-3'-dADP→3' H_2N-3'-dATP

Ribonucleotides→Deoxyribonucleotides

DNA

Fig. 39. Phosphorylations of 3'-H_2N-3'-dAR and its possible inhibitory effects as indicated by ▶. Broken lines represent inhibitions not definitely established

antibiotic strongly inhibits the biosynthesis of both DNA and RNA in these cells. Interference with nucleotide metabolism is not significant at low concentrations of the inhibitor, although with increasing concentrations there is progressive diminution in size of ribonucleotide pool (TRUMAN and KLENOW, 1968). SHIGEURA et al. (1966) observed that 3'-H_2N-3'-dATP inhibits the activity of RNA polymerase obtained from *M. lysodeikticus*. The inhibition of RNA synthesis is about eighty percent at a ratio of ATP to 3'-H_2N-3'-dATP of 5:2. Since addition of increasing amounts of the antibiotic triphosphate, in the absence of ATP, does not significantly stimulate RNA synthesis, it is possible that 3'-H_2N-3'-dATP functions in a similar way to ATP in being incorporated into RNA, but in contrast to ATP it may prevent further elongation of RNA chains due to the absence of a free hydroxyl group at the 3 position of the sugar moiety. A profound inhibition of RNA synthesis has also been observed in isolated cell nuclei or in cell-free RNA polymerase systems of Ehrlich ascites cells (TRUMAN and KLENOW, 1968). The cell-free system showed a complete blockage of the reaction even at a concentration ratio of 50:1 between ATP and $3'-H_2N-3'-dATP$. Prior to this complete inhibition there appears to be an induction period of continuing RNA synthesis that varies directly with the molar ratio of ATP to $3'-H_2N-3'-dATP$. These results also support the irreversible incorporation of a nucleoside monophosphate lacking a 3'-OH group at the 3' end of a polyribonucleotide chain growing in the usual 5' to 3' direction (Fig. 40).

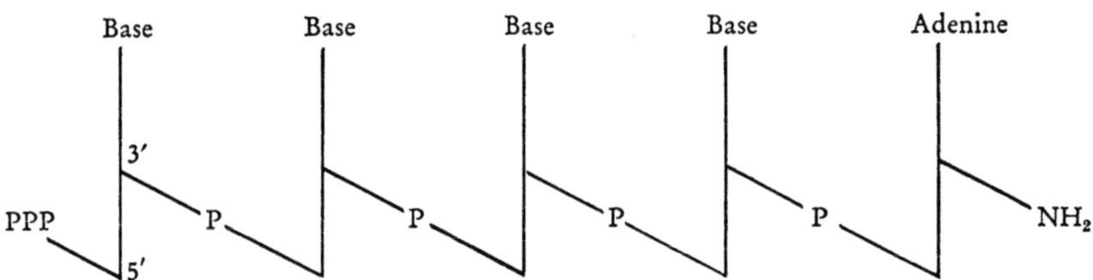

Fig. 40. Schematic diagram showing a RNA chain formed in the presence of $3'-H_2N-3'-dATP$ (TRUMAN and KLENOW, 1968)

TRUMAN and KLENOW (1968) also investigated the effect of $3'-H_2N-3'-dATP$ on DNA synthesis from deoxyribonucleoside triphosphates using a supernatant fraction from Ehrlich ascites cells extract. They failed to demonstrate inhibition of DNA synthesis even at a molar ratio of $3'-H_2N-3'-dATP$ to dATP of 1:1. Thus, it appears that the inhibition of DNA synthesis by the antibiotic proceeds through a mechanism other than what is likely for RNA synthesis. However, before this conclusion can be made it remains to be shown that 2'-deoxyribonucleotides of $3'-H_2N-3'-dAR$ are not formed in these cells. It is possible that the nucleotides of the antibiotic may cause inhibition of ribonucleotide reductases. This would limit the production of deoxyribonucleotides and thus would cause inhibition of DNA synthesis. It would be of interest to investigate this possibility with isolated enzyme systems.

Showdomycin

Showdomycin

Uridine

Showdomycin was first isolated from *Streptomyces showdoensis* by NISHIMURA et al. (1964). The antibiotic inhibits the growth of both gram-positive and gram-negative bacteria and especially the growth of *Streptococcus hemolyticus* and *Streptococcus pyrogenes* (NISHIMURA et al., 1964; NISHIMURA, 1964). Showdomycin is also a potent inhibitor of Ehrlich ascites tumor *in vivo* and HeLa cells in tissue culture (NISHIMURA et al., 1964; MATSUURA, SHIRATORI and KATAGIRI, 1964). The structure of showdomycin has been shown to be 3-β-D-ribofuranosylmaleimide (DARNALL, TOWNSEND and ROBINS, 1967; KANO et al., 1967).

The unique structure of showdomycin, its structural similarity to uridine, and its biologic properties have prompted investigations of its metabolism and effects on various enzymes, particularly those involved in uridine metabolism. It was found that the antibiotic is not a substrate for nucleoside kinase or nucleoside phosphorylase preparations of Ehrlich ascites cells (ROY-BURMAN, ROY-BURMAN and VISSER, 1968 a). The resistance of showdomycin towards enzymatic phosphorolysis may be predicted because of the chemical stability of its carbon-linked glycosyl bond. It is similar in this respect to pseudouridine which is also not acted upon by nucleosidase or nucleoside phosphorylase.

Showdomycin exhibits inhibitory effects on certain enzymes in Ehrlich ascites cell preparations which are involved in uridine and orotic acid metabolism. It inhibits UMP kinase, uridine phosphorylase, and possibly orotic acid phosphoribosyltransferase, but has no effect on the activity of uridine kinase or adenosine phosphorylase (ROY-BURMAN, ROY-BURMAN and VISSER, 1968 a). The authors also investigated the effect of showdomycin on some other enzymes. It strongly inhibits bovine liver UDP-glucose dehydrogenase, but has no effect on rabbit muscle lactic acid dehydrogenase. It is a weak inhibitor of *E. coli* RNA polymerase, but is practically inactive against *E. coli* DNA polymerase. The inhibition of the sensitive enzymes is highly pronounced when the enzymes are preincubated with the antibiotic.

The diversity in the nature of enzymatic reactions inhibited by showdomycin, its inertness as a substrate for phosphorylation reaction, and the marked sensitivity of the enzymes to the preincubation suggest possible inhibitory mechanisms which are not related to simple structural association with the enzyme at active or allosteric sites. The aglycon moiety of showdomycin is structurally related to maleimide. Maleimide derivatives are known to be active alkylating agents for sulfhydryl groups of amino acids, peptides, or proteins at physiological pH and temperature ranges (GREGORY, 1955; LESLIE, 1965; MORELL et al., 1964; MOSTELLER, RAVEL and HARDESTY, 1966). These compounds are also known to inactivate enzymes by similar alkylating action (GREGORY, 1955; MOSTELLER, RAVEL and HARDESTY, 1966). It has also been reported that under certain conditions N-substituted maleimides react with amines and amino acids (SHARPLESS and FLAVIN, 1966); however, the rate of reaction with sulfhydryl groups is much greater than that with amino groups. For example, at about pH 7 the reaction of proline with N-ethylmaleimide is only 10^{-3} that of either cysteine or glutathione (SHARPLESS and FLAVIN, 1966). Evidence that showdomycin exerts its inhibitory effect by alkylation has been provided by the selective effect of cysteine in preventing the inhibitory action of the antibiotic (ROY-BURMAN, ROY-BURMAN and VISSER, 1968 a). Preincubation of showdomycin with cysteine completely removes its inhibitory effect on UDP-glucose dehydrogenase. This protective effect of cysteine does not occur when the enzyme is preincubated with showdomycin prior to addition of cysteine. Chemical interaction of the antibiotic with cysteine has been demonstrated. This evidence together with reaction kinetics showing the uncompetitive nature of UDP-glucose dehydrogenase inhibition by showdomycin, conclusively demonstrate that the major action of showdomycin is due to its alkylating effect on enzymes. The most likely sites of alkylation are the accessible sulfhydryl groups (Fig. 41). The selective inhibition of certain enzymes involved in the metabolism of pyrimidine compounds may therefore be assumed to be related to the availability of sulfhydryl reactive sites on the sensitive enzymes. This is consistent with the inactivity of showdomycin in inhibiting *E. coli* DNA polymerase. The enzyme has been shown recently to contain only one $-$SH group whose chemical alteration has no effect on the enzyme activity (KORNBERG, 1969). It is of interest in this respect that adenosine phosphorylase is completely resistant to showdomycin, whereas uridine phosphorylase is strongly inhibited.

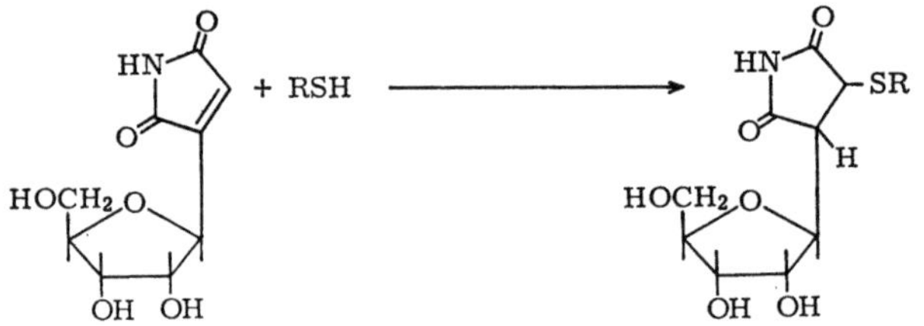

Fig. 41. Mechanism of alkylation by showdomycin as would be expected from the known addition of sulfhydryl group to the double bond of maleimide

Reports from other laboratories also support an alkylating mechanism for showdomycin. HADLER, CLAYBOURN and TSCHANG (1968) showed that showdomycin induces mitochondrial volume changes which is attributed to a chemical reaction between the antibiotic and a mitochondrial thiol group normally involved in oxidative phosphorylation. NISHIMURA and KOMATSU (1968) observed that the growth inhibitory activity of showdomycin on *E. coli* (UMEZAWA) is reversed by thiol compounds like glutathione, β-mercaptoethanol and cysteine. All the amino acids except cysteine are ineffective in reversing inhibition. The authors also reported that all the ribo- and deoxyribonucleosides have a strong reversing effect on showdomycin activity and no significant difference in effectiveness is observed among the nucleosides, while purine and pyrimidine bases, ribose, deoxyribose, nucleotides and sugar phosphates are inactive in this respect. This is of considerable interest and warrants further investigations. Assuming that showdomycin acts primarily as an alkylating agent, an explanation for the reversing effects of nucleosides possibly may be sought in their ability to induce production enzymes which are subject to showdomycin inhibition. Alternatively, showdomycin may interact preferentially with the membrane proteins because they are first exposed to the antibiotic. Conceivably, such interactions may be damaging to the cells, and nucleosides may be involved in some unknown manner in restoring the damaged processes.

Chapter 5

Conclusion

The known inhibitory effects of purine, pyrimidine and nucleoside analogues have been discussed in the preceding chapters. A general observation is that most of the analogues can participate in many reactions of their normal counterparts and may inhibit at multiple *loci*. Substrate or inhibitory activity of a particular analogue on a specific enzymic reaction is obtained mainly from studies with the isolated enzyme. In certain instances a correlation between *in vitro* and *in vivo* observations is possible. A major problem encountered, however, is the lack of cohesive and systematic studies. This is obviously a result of the use of a variety of biological systems in various laboratories interested in different aspects of inhibitor actions. The biological systems used vary from bacteria to the more complex tissues of vertebrates. Thus, the problem is not just the questionable validity of the extrapolation of *in vitro* data to living cells, but also the existence of variations in the enzymes, nucleic acids, and other cellular components of the species investigated. For these reasons, the present compilation, particularly the outline formats, should be viewed with considerable reservation. It should also be remembered that additional inhibitory effects, which are as yet unknown, may have equal or even more importance in living cells.

At this time there are few direct methods of determining an analogue's mechanism of inhibitory action in cells, tissues, or a system of tissues. To develop an understanding of its mechanism of inhibition, one has to use the means available which are based on disrupted cellular systems. Studies of this nature may be viewed as building blocks whose accumulation and interpretations could lead to knowledge of the effects of the inhibitor in living organisms. This monograph provides a background by presenting collected pieces of information derived from a wide spectrum of studies with analogues of nucleic acid components. It considers enzyme inhibition by substrate analogues at active and allosteric sites and some of the consequences of analogue incorporation into macromolecules.

This discussion has been restricted to a few selected purine, pyrimidine and nucleoside analogues. Extensive modifications among the biologically interesting analogues have resulted in an enormous number of compounds and thus it becomes a formidable task to review all of these compounds in depth in one volume. It should be noted that omission of certain compounds does not imply that they are less important. In general, those compounds which have been omitted were investigated less extensively. A larger number of analogues is listed in a recent monograph (TIMMIS, 1967). Specifically, it should be noted that the present monograph has excluded a large class of pyrimidine analogues which are primarily inhibitors of the function of folic acid. These are 2,4-diaminopyrimidines (HITCHINGS and BURCHALL, 1966),

5-arylazo derivatives of 2,4,6-triaminopyrimidines (Timmis et al., 1957; Modest, Schlein and Foley, 1957; Kawashima, 1959), and 5-arylazo derivatives of N-pyrimidyl amino acids (Roy-Burman and Sen, 1962, 1964).

Future research directed toward an understanding of the mode of action of analogues will be greatly stimulated by experimental conditions which approximate *in vivo* situations. Since most analogues require prior conversion to phosphorylated derivatives to exert inhibitory effects, cellular impermeability to nucleotides is a barrier to their application in specific inhibition studies in whole cell or tissue systems. It is of interest to note, in this connection, that methods are available which can alter permeability of bacterial cells. Gros et al. (1967) have described a technique by which *E. coli* can be made permeable to phosphorylated substrates, thus permitting the study of RNA polymerase in whole cells. The alteration in permeability of the cells is caused by plasmolysis in sucrose solution. This type of system, which more closely approximates *in vivo* conditions, could find important use in determining the effects of nucleotide analogues. For example, recent studies on the incorporation of 5,6-dihydrouridine 5'-triphosphate (H_2UTP) into RNA of plasmolyzed *E. coli* cells have shown that H_2UTP is incorporated into the fraction of RNA corresponding to transfer RNA and not into higher molecular weight RNA fractions (Roy-Burman and Visser, 1969). Earlier studies with purified *E. coli* RNA polymerase, however, could not point to this specificity of H_2UTP incorporation (Roy-Burman, Roy-Burman and Visser, 1965 a, 1967). Conceivably, cell systems with altered permeability could provide useful information concerning specific incorporation or effects of nucleotide analogues on species of RNA, on DNA synthesis, and on reactions of protein synthesis and cofactor metabolism. Future emphasis should be on supplementation of observations *in vivo* with information from systems which permit determination of individual enzyme reactions in approximately normal environmental conditions.

The ultimate aim of this field of study is to improve the chemotherapeutic applications of the analogues in viral, bacterial, parasitic and neoplastic diseases. Concepts of the biochemical mechanism of action of individual analogues have provided insights into the possible advantages of their use in combination. Potentiation of inhibitory activity by the synergistic action of two or more inhibitors has been observed. Advantages of synergism are that smaller doses of drug can be used and the emergence of resistant cells is reduced. Another potenial outcome of this research is the development of selective toxicity approaches in chemotherapy. It is conceivable that in the absence of qualitative metabolic differences between the offending growth and the host tissues, differences in enzyme specificity with regard to analogues may be utilized as a rational basis for selective effects. This is illustrated by the inhibition of mammalian DNA polymerase by ara-ATP and ara-CTP, and their inertness toward bacterial DNA polymerase. The uridine analogue, 6-aza-UR is converted to 6-aza-UMP in both bacterial and mammalian systems. However, 6-aza-UMP is further phosphorylated to di- and triphosphates in bacteria but not in mammalian tissues. An interesting example is 5-aza-CR. In *E. coli* infected with phage T4 the inhibition of DNA synthesis is the only observable effect with 5-aza-CR, whereas in uninfected cells the analogue does not inhibit DNA synthesis. The data indicates that the analogue interferes specifically with viral DNA replication. Another example which is related directly to the basis for the chemotherapeutic action of the antifolate compound, pyrimethamine, is that *Plasmodium berghei* possesses a unique dihydrofolate reductase which shows over 1000-fold greater binding of the antifolate as compared to the host enzyme (FERONE, BURCHALL and HITCHINGS, 1969).

There can be no doubt that in the future, with a greater understanding of the biochemical interrelationship between the infecting organism and the host cell, information will be available for the more skillful application of analogues in chemotherapy. Progress in comparative biochemistry of the normal and neoplastic tissues could focus attention toward subtle biochemical differences which are as yet unknown. It would then be possible to design analogues to take advantage of such differences. Although differences may reside at any point of innumerable cellular events, at this moment it appears interesting to investigate possible differences in the initiation or termination factors required for macromolecular biosynthesis, and in factors required for structural organization. Thus, from both theoretical and practical points of view there is continuing hope for the antimetabolite approach to cancer chemotherapy. It is also expected that new analogues will continue to appear from arbitrary or logical modifications of existing analogues as well as from the discovery of new antibiotics. These developments will provide tools with a wider range of selectivity for application against specific neoplastic growths. Thus, the concepts which were originated two decades ago are still valid.

References

ACKERMANN, W. W., and V. R. POTTER: Enzyme inhibition in relation to chemotherapy. Proc. Soc. exp. Biol. (N. Y.) 72, 1 (1949).

ACS, G., and E. REICH: Tubercidin and related pyrrolopyrimidine antibiotics. In: Antibiotics, Vol. I. Mechanism of Action. Eds.: D. GOTTLIEB and P. D. SHAW. New York: Springer 1967, p. 494.

— —, and M. MORI: Biological and biochemical properties of the analogue antibiotic tubercidin. Proc. nat. Acad. Sci. (Wash.) 52, 493 (1964).

AHMANN, D. L., H. F. BISEL, and R. G. HAHN: An evaluation of 5-fluorouracil in the treatment of advanced breast cancer. Mayo Clin. Proc. 42, 193 (1967).

ALLAN, P. W., H. P. SCHNEBLI, and L. L. BENNETT JR.: Conversion of 6-mercaptopurine and 6-mercaptopurine ribonucleoside to 6-methylmercaptopurine ribonucleoside in human epidermoid carcinoma No. 2 cells in culture. Biochim. biophys. Acta (Amst.) 114, 647 (1966).

ANDERSON, E. E., R. W. RUNDLES, H. R. SILBERMAN, and E. N. METZ: Allopurinol control of hyperuricosuria: a new concept in the prevention of uric acid stones. J. Urol. (Baltimore) 97, 344 (1967).

ANDERSON, P. M., and A. MEISTER: Control of *Escherichia coli* carbamylphosphate synthetase by purine and pyrimidine nucleotides. Biochemistry 5, 3164 (1966).

ANDOH, T., and E. CHARGAFF: Formation and fate of abnormal ribosomes of *E. coli* cells treated with 5-fluorouracil. Proc. nat. Acad. Sci. (Wash.) 54, 1181 (1965).

ANZAI, K., G. NAKAMURA, and S. SUZUKI: A new antibiotic, tubercidin. J. Antibiot. (Tokyo), Ser. A, 10, 201 (1957).

ARNSTEIN, H. R. V.: Mechanisms of protein biosynthesis. Brit. med. Bull. 21, 217 (1965).

ARONSON, A. I.: The effect of 5-fluorouracil on bacterial protein and ribonucleic acid synthesis. Biochim. biophys. Acta (Amst.) 49, 98 (1961).

ATKINSON, M. R., G. ECKERMANN, and J. STEPHENSON: Formation of 6-thioxanthosine 5'-phosphate from 6-mercaptopurine and from 6-thioxanthine in Ehrlich ascites-tumor cells. Biochim. biophys. Acta (Amst.) 108, 320 (1965).

ATKINSON, M. R., J. F. JACKSON, and R. K. MORTON: Substrate specific and inhibition of nicotinamide mononucleotide-adenyltransferase of liver nuclei: possible mechanism of effect of 6-mercaptopurine on tumour growth. Nature 192, 946 (1961).

—, R. K. MORTON, and A. W. MURRAY: Inhibition of inosine 5′-phosphate dehydrogenase from Ehrlich ascites-tumour cells by 6-thioinosine 5′-phosphate. Biochem. J. 89, 167 (1963).

— — — Inhibition of adenylosuccinate synthetase and adenylosuccinate lyase from Ehrlich ascites-tumour cells by 6-thioinosine 5′-phosphate. Biochem. J. 92, 398 (1964).

—, and A. W. MURRAY: Inhibition of purine phosphoribosyltransferase of Ehrlich ascites-tumour cells by 6-mercaptopurine. Biochem. J. 94, 64 (1965).

BAKER, B. R., R. E. SCHAUB, and H. M. KISSMAN: Puromycin. Synthetic studies. XV. 3′-Amino-3′-deoxyadenosine. J. Amer. chem. Soc. 77, 5911 (1955).

BAUGNET-MAHIEU, L., and R. GOUTIER: Mechanisms responsible for the low incorporation into DNA of the thymidine analogue, 5-iodo-2′-deoxyuridine. Biochem. Pharmacol. 17, 1017 (1968).

BAUTZ, E. K. F.: RNA synthesis-mechanism of genetic transcription. In: Molecular Genetics, Part 2. Ed.: J. H. TAYLOR. New York: Academic Press 1967, p. 213.

BELTZ, R. E., and D. W. VISSER: Studies on the action of thymidine analogues. J. biol. Chem. 226, 1035 (1957).

BEN-ISHAI, R., and B. E. VOLCANI: Dependence of protein synthesis on ribonucleic acid formation in a thymine-requiring mutant of *Escherichia coli*. Biochim. biophys. Acta (Amst.) 21, 265 (1956).

BENNETT, L. L., JR., R. W. BROCKMAN, H. P. SCHNEBLI, S. CHUMLEY, G. J. DIXON, F. M. SCHABEL, JR., E. A. DULMADGE, H. E. SKIPPER, J. A. MONTGOMERY, and H. J. THOMAS: Activity and mechanism of action of 6-methylthiopurine ribonucleoside in cancer cells resistant to 6-mercaptopurine. Nature 205, 1276 (1965).

—, H. P. SCHNEBLI, M. H. VAIL, P. W. ALLAN, and J. A. MONTGOMERY: Purine ribonucleoside kinase activity and resistance to some analogs of adenosine. Molec. Pharmacol. 2, 432 (1966).

—, L. SIMPSON, J. GOLDEN, and T. L. BARKER: The primary site of inhibition by 6-mercaptopurine on the purine biosynthetic pathway in some tumors *in vivo*. Cancer Res. 23, 1574 (1963).

BEN-PORAT, T., M. BROWN, and A. S. KAPLAN: Effect of 1-β-D-arabinofuranosylcytosine on DNA synthesis. II. In rabbit kidney cells infected with herpes viruses. Molec. Pharmacol. 4, 139 (1968).

—, A. S. KAPLAN, and R. W. TENNANT: Effect of 5-fluorouracil on the multiplication of a virulent virus (pseudorabies) and on oncogenic virus (polyoma). Virology 32, 445 (1967).

BENTLEY, H. R., K. G. CUNNINGHAM, and F. S. SPRING: Structure of cordycepin. J. chem. Soc. 2301 (1951).

BESSMAN, M. J., I. R. LEHMAN, J. ALDER, S. B. ZIMMERMAN, E. S. SIMMS, and A. KORNBERG: Enzymatic synthesis of deoxyribonucleic acid. III. The incorporation of pyrimidine and purine analogues into deoxyribonucleic acid. Proc. nat. Acad. Sci. (Wash.) 44, 633 (1958).

BHUYAN, B. K., H. E. RENIS, and C. G. SMITH: A collagen plate assay for cytotoxic agents. II. Biological studies. Cancer Res. 22, 1131 (1962).

BIEBER, A. L., and A. C. SARTORELLI: The metabolism of thioguanine in purine analog-resistant cells. Cancer Res. 24, 1210 (1964).

BIEBER, S., L. S. DIETRICH, G. B. ELION, G. H. HITCHINGS, and D. S. MARTIN: The incorporation of 6-mercaptopurine-^{35}S into the nucleic acids of sensitive and nonsensitive transplantable mouse tumors. Cancer Res. 21, 228 (1961).

—, G. B. ELION, H. C. NATHAN, and G. H. HITCHINGS: The inhibition of mammary adenocarcinoma 755 by analogs of uracil. Proc. Amer. Assoc. Cancer Res. 2, 188 (1957).

BIRNIE, G. D., H. KROGER, and C. HEIDELBERGER: Studies of fluorinated pyrimidines. XVIII. The degradation of 5-fluoro-2′-deoxyuridine and related compounds by nucleoside phosphorylase. Biochemistry 2, 566 (1963).

BLAKLEY, R. L.: The biosynthesis of thymidylic acid. IV. Further studies on thymidylate synthetase. J. biol. Chem. 238, 2113 (1963).

—, and E. VITOLS: The control of nucleotide biosynthesis. Ann. Rev. Biochem. 37, 201 (1968).

BLOCH, A., M. T. HAKALA, E. MIHICH, and C. A. NICHOL: Studies on the biological activity and mode of action of 7-deazinosine. Proc. Amer. Assoc. Cancer Res. **5**, 6 (1964).

—, R. J. LEONARD, and C. A. NICHOL: On the mode of action of 7-deazaadenosine (tubercidin). Biochim. biophys. Acta (Amst.) **138**, 10 (1967).

BRANSTER, M. V., and R. K. MORTON: Comparative rates of synthesis of diphosphopyridine nucleotide by normal and tumour tissue from mouse mammary gland: studies with isolated nuclei. Biochem. J. **63**, 640 (1956).

BRESNICK, E.: The metabolism *in vitro* of antitumor imidazolyl derivatives of mercaptopurines. Fed. Proc. **18**, 371 (1959).

—, and U. B. THOMPSON: Properties of deoxythymidine kinase partially purified from animal tumors. J. biol. Chem. **240**, 3967 (1965).

—, and S. S. WILLIAMS: Effects of 5-trifluoromethyldeoxyuridine upon deoxythymidine kinase. Biochem. Pharmacol. **16**, 503 (1967).

BRIDGER, W. A., and L. H. COHEN: The mechanism of inhibition of adenylosuccinate lyase by 6-mercaptopurine nucleotide (thioinosinate). Biochim. biophys. Acta (Amst.) **73**, 514 (1963).

BRINDLE, S. A., N. A. GIUFFRE, B. J. AMREIN, R. C. MILLONIG, and D. PERLMAN: Antibiotic sensitivity of Ehrlich ascites cells grown in tissue culture. Antimicrobial agents and chemotherapy, 1961, p. 159.

BRINK, J. J., and G. A. LEPAGE: Metabolic effects of 9-D-arabinosylpurines in ascites tumor cells. Cancer Res. **24**, 312 (1964).

— — Metabolism and distribution of 9-β-D-arabinofuranosyladenine in mouse tissues. Cancer Res. **24**, 1042 (1964 a).

— — 9-β-D-arabinofuranosyladenine as an inhibitor of metabolism in normal and neoplastic cells. Canad. J. Biochem. **43**, 1 (1965).

BROCKMAN, R. W.: Mechanisms of resistance to anticancer agents. Advanc. Cancer Res. **7**, 129 (1963).

— Metabolism and mechanisms of action of purine analogues. In: Exploitable molecular mechanisms and cancer. Published for the University of Texas M. D. Anderson Hospital and Tumor Institute at Houston. Baltimore: The Williams and Wilkins Co. 1969, p. 435.

—, and E. P. ANDERSON: Biochemistry of cancer (metabolic aspects). Ann. Rev. Biochem. **32**, 463 (1963).

—, L. L. BENNETT JR., M. S. SIMPSON, A. R. WILSON, J. R. THOMSON, and H. E. SKIPPER: A mechanism of resistance to 8-azaguanine. II. Studies with experimental neoplasms. Cancer Res. **19**, 856 (1959 a).

—, and S. CHUMLEY: Inhibition of formylglycinamide ribonucleotide synthesis in neoplastic cells by purines and analogs. Biochim. biophys. Acta (Amst.) **95**, 365 (1965).

—, C. SPARKS, D. J. HUTCHISON, and H. E. SKIPPER: A mechanism of resistance to 8-azaguanine. I. Microbiological studies on the metabolism of purines and 8-azapurines. Cancer Res. **19**, 177 (1959).

—, M. C. SPARKS, and M. S. SIMPSON: A comparison of the metabolism of purines and purine analogs by susceptible and drug-resistant bacterial and neoplastic cells. Biochim. biophys. Acta (Amst.) **26**, 671 (1957).

BROX, L. W., and A. HAMPTON: Inactivation of guanosine-5'phosphate reductase by 6-chloro-, 6-mercapto-, and 2-amino-6-mercapto-9-β-D-ribofuranosyl 5'-phosphates. Biochemistry **7**, 398 (1968).

BRUEMMER, N. C., J. F. HOLLAND, and P. R. SHEEHE: Drug effects on a target metabolic pathway and on mouse tumor growth: Azauridine and decarboxylation of orotic acid-7-^{14}C. Cancer Res. **22**, 113 (1962).

BUCHANAN, J. M., and S. C. HARTMAN: Enzymatic reactions in the synthesis of the purines. Advanc. Enzymol. **21**, 199 (1959).

BUJARD, H., and C. HEIDELBERGER: Fluorinated pyrimidines. XXVII. Attempts to determine transcription errors during the formation of fluorouracil-containing messenger ribonucleic acid. Biochemistry **5**, 3339 (1966).

BURCHALL, J. J., and G. H. HITCHINGS: Inhibitor binding analysis of dihydrofolate reductases from various species. Mol. Pharmacol. **1**, 126 (1965).

BURCHALL, J. J., R. FERONE, and G. H. HITCHINGS: Antibacterial chemotherapy. Ann. Rev. Pharmacol. **5**, 53 (1965).

BUSSARD, A., S. NAONO, F. GROS, and J. MONOD: Effects of an analog of uracil on the properties of an enzymatic protein synthesized in its presence. C. R. Acad. Sci. (Paris) **250**, 4049 (1960).

BUTHALA, D. A.: Cell culture studies on antiviral agents. I. Action of cytosine arabinoside and some comparisons with 5-iodo-2-deoxyuridine. Proc. Soc. exp. Biol. (N. Y.) **115**, 69 (1964).

CALABRESI, P.: Current status of clinical investigations with 6-azauridine, 5-iodo-2'-deoxyuridine, and related derivatives. Cancer Res. **23**, 1260 (1963).

—, S. S. CARDOSO, S. C. FINCH, M. M. KLIGERMAN, C. F. VON ESSEN, M. Y. CHU, and A. D. WELCH: Initial clinical studies with 5-iodo-2'-deoxyuridine. Cancer Res. **21**, 550 (1961).

—, and R. W. TURNER: Beneficial effects of triacetylazauridine in psoriasis and mycosis fungoides. Ann. Int. Med. **64**, 352 (1966).

— —, and E. LEFKOWITZ: Beneficial effects of azauridine in psoriasis and mycosis fungoides. Proc. Amer. Ass. Cancer Res. **5**, 10 (1964).

—, and A. D. WELCH: Chemotherapy of neoplastic diseases. In: The Pharmacological Basis of Theurapeutics. Eds.: L. S. GOODMAN and A. GOLDIN. New York: Macmillan Co. 1965, p. 1345.

CALDWELL, I. C., J. F. HENDERSON, and A. R. P. PATERSON: The metabolism of formycin by the Ehrlich ascites carcinoma and a resistant subline. Proc. Amer. Ass. Cancer Res. **7**, 11 (1966 a).

— — — The enzymatic formation of 6-(methylmercapto) purine ribonucleoside 5'-phosphate. Canad. J. Biochem. **44**, 229 (1966).

— — — Resistance to purine ribonucleoside analogues in ascites tumor. Canad. J. Biochem. **45**, 735 (1967).

CALNE, R. Y.: Inhibition of the rejection of renal homografts in dogs by purine analogues. Transplant. Bull. **28**, 65 (1961).

—, G. W. ALEXANDRE, and J. E. MURRAY: A study of the effects of drugs in prolonging survival of homologous renal transplants in dogs. Ann. N. Y. Acad. Sci. **99**, 743 (1962).

CAMERMAN, N., and J. TROTTER: 5-Iodo-2'-deoxyuridine: relation of structure to its antiviral activity. Science **144**, 1348 (1964).

CAMIENER, G. W., and C. G. SMITH: Studies on the enzymatic deamination of cytosine arabinoside. I. Enzyme distribution and species specificity. Biochem. Pharmacol. **14**, 1405 (1965).

CARBON, J. A.: The inhibition of polynucleotide phosphorylase by 6-mercaptopurine riboside 5'-diphosphate. Biochem. biophys. Res. Commun. **7**, 366 (1962).

CARDEILHAC, P. T., and S. S. COHEN: Some metabolic properties of nucleotides of 1-β-D-arabinofuranosylcytosine. Cancer Res. **24**, 1595 (1964).

CARDOSO, S. S., P. CALABRESI, and R. E. HANDSCHUMACHER: Alterations in human pyrimidine metabolism as a result of therapy with 6-azauridine. Cancer Res. **21**, 1551 (1961).

CAREY, R. W., and R. R. ELLISON: Continuous cytosine arabinoside infusions in patients with neoplastic disease. Clin. Res. **13**, 337 (1965).

CASOLA, L., R. LIM, R. E. DAVIS, and B. W. AGRANOFF: Behavioral and biochemical effects of intracranial injection of cytosine arabinoside in goldfish. Proc. nat. Acad. Sci. (Wash.) **60**, 1389 (1968).

CHAMPE, S. P., and S. BENZER: Reversal of mutant phenotypes by 5-fluorouracil: An approach to nucleotide sequences in messenger-RNA. Proc. nat. Acad. Sci. (Wash.) **48**, 532 (1962).

CHANG, P., and B. LYTHGOE: Synthesis of purine nucleosides. XXVII. 1,2,3,5-tetraacetyl-D-xylofuranose and the D-xylofuranosides of thiophylline and adenine. J. chem. Soc., 1992 (1950).

CHASSY, B. M., and R. J. SUHADOLNIK: Nucleoside antibiotics. II. Biochemical tools for studying the structural requirements for interaction at the catalytic and regulatory sites of ribonucleotide reductase from *Escherichia coli*. J. biol. Chem. **243**, 3538 (1968).

CHU, M. Y., and G. A. FISCHER: A proposed mechanism of action of 1-β-D-arabinofuranosyl-cytosine as an inhibitor of the growth of leukemic cells. Biochem. Pharmacol. **11**, 423 (1962).

CHU, M. Y., and G. A. FISCHER: Comparative studies of leukemic cells sensitive and resistant to cytosine arabinoside. Biochem. Pharmacol. **14**, 333 (1965).

— — The incorporation of ^{3}H-cytosine arabinoside and its effect on murine leukemic cells. Biochem. Pharmacol. **17**, 753 (1968 a).

— — Effect of cytosine arabinoside on the cell viability and uptake of deoxypyrimidine nucleosides in L5178Y cells. Biochem. Pharmacol. **17**, 741 (1968).

CIHAK, A., and F. SORM: Biochemical effects and metabolic transformation of 5-azacytidine in *Escherichia coli.* Coll. Czech. Chem. Commun. **30**, 2091 (1965 a).

— — Interaction of 5-azauracil with uridine phosphorylase in a cell-free extract of mouse liver. Coll. Czech. Chem. Commun. **30**, 324 (1965).

— — Metabolic transformations of 5-azaorotate: Cause of marked inhibition of orotidine 5′-phosphate decarboxylase. Biochim. biophys. Acta (Amst.) **149**, 314 (1967).

—, R. TYKVA, and F. SORM: Incorporation of 5-azacytidine-4-^{14}C and of cytidine-^{3}H into ribonucleic acids of Ehrlich ascites tumor cells. Coll. Czech. Chem. Commun. **31**, 3015 (1966).

—, J. VESELY, and F. SORM: Complete inhibition by 5-azacytidine of hormonal induction of tryptophan pyrrolase. Biochim. biophys. Acta (Amst.) **134**, 486 (1967).

CODDINGTON, A.: Some substrates and inhibitors of adenosine deaminase. Biochim. biophys. Acta (Amst.) **99**, 442 (1965).

COHEN, L. H., and R. E. PARKS JR.: Inhibition and activation of adenylosuccinic synthetase by 8-azaguanosine triphosphate. Canad. J. Biochem. **41**, 1495 (1963).

COHEN, S. S.: Introduction to the biochemistry of D-arabinosyl nucleosides. Progr. in Nucl. Acid. Res. molec. Biol. **5**, 1 (1966).

—, J. G. FLAKS, H. D. BARNER, M. R. LOEB, and J. LICHTENSTEIN: The mode of action of 5-fluorouracil and its derivatives. Proc. nat. Acad. Sci. (Wash.) **44**, 1004 (1958).

CONN, H. O., W. A. CREASEY, and P. CALABRESI: Effect of 6-azauridine on plasma cell tumors of mice: correlation of antitumor effect with inhibition of orotic acid metabolism. Cancer Res. **27**, 618 (1967).

COOK, J. L., and M. VIBERT: The utilization of purines and their ribosyl derivatives for the formation of adenosine triphosphate and guanosine triphosphate in the rabbit reticulocyte. J. biol. Chem. **241**, 158 (1966).

CORY, J. G., R. J. SUHADOLNIK, B. RESNICK, and M. A. RICH: Incorporation of cordycepin (3′-deoxyadenosine) into ribonucleic acid and deoxyribonucleic acid of human tumor cells. Biochim. Biophys. Acta (Amst.) **103**, 646 (1965).

CREASEY, W. A., M. E. FINK, R. E. HANDSCHUMACHER, and P. CALABRESI: Clinical and pharmacological studies with 2′,3′,5′-triacetyl-6-azauridine. Cancer Res. **23**, 444 (1963).

—, B. J. PAPAC, M. E. MARKIW, P. CALABRESI, and A. D. WELCH: Biochemical and pharmacological studies with 1-β-D-arabinofuranosylcytosine in man. Biochem. Pharmacol. **15**, 1417 (1966).

CROSBIE, G. W.: Biosynthesis of pyrimidine nucleotides. In: The Nucleic Acids, Vol. 3. Eds.: E. CHARGAFF and J. N. DAVIDSON. New York: Academic Press 1960, p. 323.

CUNNINGHAM, K. G., S. A. HUTCHINSON, W. MASON, and F. S. SPRING: Cordycepin, a metabolic product from cultures of *Cordyceps militaris.* I. Isolation and characterization. J. chem. Soc. 2299 (1951).

DARNALL, K. R., L. B. TOWNSEND, and R. K. ROBINS: The structure of showdomycin, a novel carbon-linked nucleoside antibiotic related to uridine. Proc. nat. Acad. Sci. (Wash.) **57**, 548 (1966).

DEINHARDT, F.: Antiviral substances, Section III. Nucleosides: Discussion. Ann. N. Y. Acad. Sci. **130**, 218 (1965).

DELAMORE, I. W., and W. H. PRUSOFF: Effect of 5-iodo-2′-deoxyuridine on the biosynthesis of phosphorylated derivatives of thymidine. Biochem. Pharmacol. **11**, 101 (1962).

DJORDJEVIC, B., and W. SZYBALSKI: Genetics of human cell lines. III. Incorporation of 5-bromo- and 5-iododeoxyuridine into the deoxyribonucleic acid of human cells and its effect on radiation sensitivity. J. exp. Med. **112**, 509 (1960).

DOERING, A., J. KELLER, and S. S. COHEN: Some effects of D-arabinosylnucleosides on polymer synthesis in mouse fibroblasts. Cancer Res. **26**, 2444 (1966).

DOLLINGER, M. R., J. H. BURCHENAL, W. KRIS, and J. J. FOX: Analogs of 1β-D-arabino-furanosylcytosine. Studies on mechanisms of action in Burkitts' cell culture and mouse leukemia, and *in vitro* deamination studies. Biochem. Pharmacol. **16**, 689 (1967).

DOSKOCIL, J., V. PACES, and F. SORM: Inhibition of protein synthesis by 5-azacytidine in *Escherichia coli*. Biochim. Biophys. Acta (Amst.) **145**, 771 (1967).

—, and F. SORM: The action of 5-azacytidine on bacteria infected with bacteriophage T_4. Biochim. Biophys. Acta (Amst.) **145**, 780 (1967).

DUNCAN, R. E., and P. S. WOODS: Some cytological aspects of antagonism in synthesis of nucleic acid. Chromosoma **6**, 45 (1953).

DUNN, D. B., and J. D. SMITH: The occurrence of 6-methylaminopurine in deoxyribonucleic acids. Biochem. J. **68**, 627 (1958).

DURHAM, J. P., and D. H. IVES: The phosphorylation of cytosine arabinoside by deoxycytidine kinase. Fed. Proc. **27**, 538 (1968).

DUSCHINSKY, R., E. PLEVEN, and C. HEIDELBERGER: The synthesis of 5-fluoropyrimidines. J. Amer. chem. Soc. **79**, 4559 (1957).

DUTTON, R. W., A. H. DUTTON, and J. H. VAUGHAN: The effect of 5-bromouracil deoxyriboside on the synthesis of antibody *in vitro*. Biochem. J. **75**, 230 (1960).

DUVALL, L. R.: Tubercidin. Cancer Chemotherapy Rept. **30**, 61 (1963).

EDLIN, G.: Amino acid regulation of bacteriophage RNA synthesis. J. molec. Biol. **12**, 356 (1965).

— Gene regulation during bacteriophage T_4 development. I. Phenotypic reversion of T_4 amber mutants by 5-fluorouracil. J. molec. Biol. **12**, 363 (1965 a).

EGGERS, H. J., and I. TAMM: Antiviral chemotherapy. Ann. Rev. Pharmacol. **6**, 231 (1966).

EIDINOFF, M. L., J. E. KNOLL, B. J. MARANO, and D. KLEIN: Pyrimidine studies. II. Effect of 5-bromodeoxyuridine and related nucleosides on incorporation of precursors into nucleic acid pyrimidines. Cancer Res. **19**, 738 (1959).

ELION, G. B.: Biochemistry and pharmacology of purine analogs. Fed. Proc. **26**, 898 (1967).

—, S. BIEBER, and G. H. HITCHINGS: The fate of 6-mercaptopurine in mice. Ann. N. Y. Acad. Sci. **60**, 297 (1954).

— —, H. NATHAN, and G. H. HITCHINGS: Uracil antagonism and inhibition of mammary adenocarcinoma 755. Cancer Res. **18**, 802 (1958).

—, E. BURGI, and G. H. HITCHINGS: Studies on condensed pyrimidine systems. IX. The synthesis of some 6-substituted purines. J. Amer. chem. Soc. **74**, 411 (1952).

—, S. CALLAHAN, S. BIEBER, G. H. HITCHINGS, and R. W. RUNDLES: A summary of investigations with 6[(1-methyl-4-nitro-5-imidazolyl) thio] purine (B. W. 57-322). Cancer Chemother. Rept. **14**, 93 (1961).

— —, R. W. RUNDLES, and G. H. HITCHINGS: Relationship between metabolic fates and antitumor activities of thiopurines. Cancer Res. **23**, 1207 (1963).

—, S. W. CALLAHAN, G. H. HITCHINGS, R. W. RUNDLES, and J. LASZLO: Experimental, clinical, and metabolic studies of thiopurines. Cancer Chemother. Rept. **16**, 197 (1962).

— —, H. NATHAN, S. BIEBER, R. W. RUNDLES, and G. H. HITCHINGS: Potentiation by inhibition of drug degradation: 6-substituted purines and xanthine oxidase. Biochem. Pharmacol. **12**, 85 (1963 a).

—, and G. H. HITCHINGS: Metabolic basis for the actions of analogs of purines and pyrimidines. In: Advances in Chemotherapy, Vol. 2. Eds.: A. GOLDIN, F. HAWKING, and R. J. SCHNITZER. New York: Academic Press 1965, p. 91.

—, A. KOVENSKY, G. H. HITCHINGS, E. METZ, and R. W. RUNDLES: Metabolic studies of allopurinol, an inhibitor of xanthine oxidase. Biochem. Pharmacol. **15**, 863 (1966).

—, S. MUELLER, and G. H. HITCHINGS: Studies on condensed pyrimidine systems. XXI. The isolation and synthesis of 6-mercapto-2,8-purinediol (6-thiouric acid). J. Amer. chem. Soc. **81**, 3042 (1959).

—, R. W. RUNDLES, and G. H. HITCHINGS: The route from 6-mercaptopurine to sulfate in man. Proc. Amer. Ass. Cancer Res. **5**, 17 (1964).

—, T. J. TAYLOR, and G. H. HITCHINGS: Binding of substrates and inhibitors to xanthine oxidases from different species. Abstracts, 6th Int. Congr. of Biochem. **4**, 305 (1964).

ELLIS, D. B., and G. A. LEPAGE: Biochemical studies of resistance to 6-thioguanine. Cancer Res. **23**, 436 (1963).

ELLIS, D. B., and G. A. LEPAGE: Some inhibitory effects of 9-β-D-xylofuranosyladenine, an adenosine analog, on nucleotide metabolism in ascites tumor cells. Molec. Pharmacol. 1, 231 (1965 a).
— — The effect of β-xylosyladenine on the formation of 5-phosphoribosyl-1-pyrophosphate by cell-free extracts of TA3 ascites cells. Can. J. Biochem. 43, 617 (1965).
— — Metabolic fate 9-β-D-xylofuranosyladenine in mice bearing susceptible tumor cells. Cancer Res. 26, 893 (1966).
ERIKSON, R. L., and W. SZYBALSKI: Molecular radiobiology of human cell lines. III. Radiation-sensitizing properties of 5-iododeoxyuridine. Cancer Res. 23, 122 (1963).
EVANS, J. S., and G. D. MENGEL: The reversal of cytosine arabinoside activity *in vivo* by deoxycytidine. Biochem. Pharmacol. 13, 989 (1964).
—, E. A. MUSSER, L. BOSTWICK, and G. D. MENGEL: The effect of 1-β-D-arabinofuranosylcytosine hydrochloride on murine neoplams. Cancer Res. 24, 1285 (1964).
— —, G. D. MENGEL, K. R. FORSBLAD, and J. H. HUNTER: Antitumor activity of 1-β-D-arabinofuranosylcytosine hydrochloride. Proc. Soc. exp. Biol. (N. Y.) 106, 350 (1961).
FALLON, H. J., E. FREI, III, and E. J. FREIREICH: Correlations of the biochemical and clinical effects of 6-azauridine in patients with leukemia. Amer. J. Med. 33, 526 (1962).
FEIGELSON, P., J. D. DAVIDSON, and R. K. ROBINS: Pyrazolopyrimidines as inhibitors and substrates of xanthine oxidase. J. biol. Chem. 226, 993 (1957).
FERONE, R., J. J. BURCHALL, and G. H. HITCHINGS: *Plasmodium berghei* dihydrofolate reductase. Isolation, properties, and inhibition by antifolates. Molec. Pharmacol. 5, 49 (1969).
FISCHER, D. S., E. P. CASSIDY, and A. D. WELCH: Immunosuppression by pyrimidine nucleoside analogs. Biochem. Pharmacol. 15, 1013 (1966).
FLAKS, J. G., and L. N. LUKENS: The enzymes of purine nucleotide synthesis *de novo*. Meth. in Enzymol. 6, 52 (1963).
FOX, J. J., K. A. WATANABE, and A. BLOCH: Nucleoside antibiotics. Progr. in Nucl. Acid Res. molec. Biol. 5, 251 (1966).
FOX, M., and L. G. LAJTHA: Continuous irradiation of P388F lymphoma *in vitro* sensitization by 5-iodo-2'-deoxyuridine. Intern. J. Radiat. Biol. 12, 251 (1967).
FREDERIKSEN, S., and H. KLENOW: Differential inhibition by 3'-deATP of nuclear and cytoplasmic RNA fractions of Ehrlich ascites tumor cells *in vitro*. Biochem. biophys. Res. Commun. 17, 165 (1964).
—, H. MALLING, and H. KLENOW: Isolation of 3'-deoxyadenosine (cordycepin) from the liquid medium of *Cordycepsmilitaris* (L. ex Fr.) Link. Biochim. biophys. Acta (Amst.) 95, 189 (1965).
FREI, E.: Chemotherapy of acute leukemia. In: Cancer Chemotherapy. Eds.: I. BRODSKY, S. B. KAHN, and J. H. MOYER. New York: Grune & Stratton 1967, p. 185.
— Effect of cancer chemotherapeutic agents on normal tissues in man. Fed. Proc. 26, 918 (1967 a).
FUCIK, V., Z. SORMOVA, and F. SORM: Effect of 5-azacytidine on the root meristem of *Vicia faba*. Biologia Plantarum 7, 58 (1965).
—, S. ZADRAZIL, Z. SORMOVA, and F. SORM: Mutagenic effects of 5-azacytidine in bacteria. Coll. Czech. Chem. Commun. 30, 2883 (1965).
FURTH, J. J., and S. S. COHEN: Inhibition of mammalian DNA polymerase by the 5'-triphosphate of 9-β-D-arabinofuranosyladenine. Cancer Res. 27, 1528 (1967).
GERBER, N. N., and H. A. LECHEVALIER: 3'-Amino-3'-deoxyadenosine, an antitumor agent from *Helminthosporium* sp. J. Org. Chem. 27, 1731 (1962).
GOLDBERG, I. H., and M. RABINOWITZ: Inhibition of RNA nucleotidyltransferase by 6-azauridine triphosphate. Biochim. biophys. Acta (Amst.) 72, 116 (1963).
GOLDBERG, N. D., J. L. DAHL, and R. E. PARKS JR.: Uridine diphosphate glucose dehydrogenase. pH dependence of the reactions with the 5-fluorouracil and 6-azauracil analogues of uridine diphosphate glucose. J. Biol. Chem. 238, 3109 (1963).
GONTCHAROFF, M., and D. MAZIA: Developmental consequences of introduction of bromouracil into the DNA of sea urchin embryos during early division stages. Exp. Cell Res. 46, 315 (1967).
GORDON, M. P., and M. STAEHELIN: Studies on the incorporation of 5-fluorouracil into a virus nucleic acid. Biochim. biophys. Acta (Amst.) 36, 351 (1959).

GOTTLIEB, D., and P. D. SHAW (eds.): Antibiotics, Vol. 1. Mechanism of Action. New York: Springer 1967.

GOTTSCHLING, H., and C. HEIDELBERGER: Fluorinated pyrimidines. XIX. Some biological effects of 5-trifluoromethyluracil and 5-trifluoromethyl-2'-deoxyuridine on *Escherichia coli* and bacteriophage T4B. J. molec. Biol. 7, 541 (1963).

GOULIAN, M., A. KORNBERG, and R. L. SINSHEIMER: Enzymatic synthesis of DNA. XXIV. Synthesis of infectious phage ΦX 174 DNA. Proc. nat. Acad. Sci. (Wash.) 58, 2321 (1967).

GOZ, B., and W. H. PRUSOFF: The ability of phage containing 5-iodo-2'-deoxyuridine-substituted deoxyribonucleic acid to induce enzymes. J. biol. Chem. 243, 4750 (1968).

GRANAT, P., W. A. CREASEY, P. CALABRESI, and R. E. HANDSCHUMACHER: Investigations with 5-azaorotic acid, an inhibitor of the biosynthesis of pyrimidines *de novo*. Clin. Pharmacol. Ther. 6, 436 (1965).

GRAV, H. J.: Nucleotides and nucleotide metabolism. In: Methods in Cancer Research, Vol. 3. Ed.: H. BUSCH. New York: Academic Press 1967, p. 243.

GREGORY, J. D.: The stability of N-ethylmaleimide and its reaction with sulfhydryl groups. J. Amer. chem. Soc. 77, 3922 (1955).

GRESSEL, J., and E. GALUN: Effect of 5-fluorouracil on the soluble RNA of Trichoderma. Biochem. biophys. Res. Commun. 24, 162 (1966).

GRINDEY, G. B., L. D. SASLAW, and V. S. WARAVDEKAR: Effects of uracil derivatives on phosphorylation of arabinosylcytosine. Molec. Pharmacol. 4, 96 (1968).

GROS, F., J. GALLANT, R. WEISBERG, and M. CASHEL: Decryptification of RNA polymerase in whole cells of *Escherichia coli*. J. molec. Biol. 25, 555 (1967).

—, and S. NAONO: Bacterial synthesis of "modified" enzymes in the presence of a pyrimidine analogue. In: Protein Biosynthesis. Ed.: R. J. C. HARRIS. London: Academic Press 1961, p. 195.

GRUNBERGER, D., A. HOLY, and F. SORM: Synthesis and coding properties of 8-azaguanosine-containing trinucleoside diphosphates. Biochim. biophys. Acta (Amst.) 161, 147 (1968).

—, L. MEISSNER, A. HOLY, and F. SORM: The effect of polymers and trinucleoside diphosphates containing 8-azaguanine upon the binding of ^{14}C-valine-sRNA to ribosomes. Biochim. biophys. Acta (Amst.) 119, 432 (1966).

GRUNBERG-MANAGO, M., and A. M. MICHELSON: Polynucleotide analogues. II. Stimulation of amino acid incorporation by polynucleotide analogues. Biochim. biophys. Acta (Amst.) 80, 431 (1964).

GUARINO, A. J.: Cordycepin. In: Antibiotics, Vol. I. Mechanism of Action. Eds.: D. GOTTLIEB and P. D. SHAW. New York: Springer 1967, p. 468.

HABERMAN, V.: The effect of 6-azauracil on microorganisms inhibited by chloramphenicol. Biochim. biophys. Acta (Amst.) 49, 204 (1961).

HADLER, H. I., B. E. CLAYBOURN, and T. P. TSCHANG: Mitochondrial volume changes induced by the antibiotic showdomycin. Biochem. biophys. Res. Commun. 31, 25 (1968).

HALL, T. C., M. J. KRANT, J. B. LLOYD, W. B. PATTERSON, A. ISHIBARA, K. G. POTEE, T. O. LOVINA, and J. M. MULLEN: Treatment of localized inoperable neoplasms with intra-arterial infusions of 8-azaguanine. Cancer 15, 1156 (1962).

HAMPTON, A.: Reactions of ribonucleotide derivatives of purine analogues at the catalytic site of inosine 5'-phosphate dehydrogenase. J. biol. Chem. 238, 3068 (1963).

HANDSCHUHMACHER, R. E.: Azauracil riboside and ribonucleotides: Isolation and chemical synthesis. J. biol. Chem. 235, 764 (1960).

— 5-Azaorotic acid and related inhibitors of the synthesis *de novo* of pyrimidine nucleotides. Cancer Res. 23, 634 (1963).

—, P. CALABRESI, A. D. WELCH, V. BONO, H. FALLON, and E. FREI, III: Summary of current information on 6-azauridine. Cancer Chemother. Rept. 21, 1 (1962).

—, and C. A. PASTERNAK: Inhibition of orotidylic acid decarboxylase, a primary site of carcinostasis by 6-azauracil. Biochim. biophys. Acta (Amst.) 30, 451 (1958).

—, and A. D. WELCH: Microbial studies of 6-azauracil, an antagonist of uracil. Cancer Res. 16, 965 (1956).

HANNA, C., and K. P. WILKINSON: Effect of iodoxuridine on the uptake of tritium-labeled thymidine in the rabbit cornea infected with herpes simplex. Exptl. Eye Res. 4, 31 (1965).

HARTMANN, K.-U., and C. HEIDELBERGER: Studies on fluorinated pyrimidines. XIII. Inhibition of thymidylate synthetase. J. biol. Chem. **236**, 3006 (1961).

HAUT, W. F., and J. H. TAYLOR: Studies of bromouracil deoxyriboside substitution in DNA of bean roots *(Vicia faba)*. J. molec. Biol. **26**, 389 (1967).

HAWKING, F.: Chemotherapy of trypanosomiasis. In: Experimental Chemotherapy, Vol. 1. Eds.: R. J. SCHNITZER and F. HAWKING. New York: Academic Press 1963, p. 203.

HEIDELBERGER, C.: Fluorinated pyrimidines. Progr. Nucl. Acid. Res. molec. Biol. **4**, 1 (1965).

— Cancer chemotherapy with purine and pyrimidine analogues. Ann. Rev. Pharmacol. **7**, 101 (1967).

—, and S. W. ANDERSON: Fluorinated pyrimidines. XXI. The tumor inhibitory activity of 5-trifluoromethyl-2'-deoxyuridine. Cancer Res. **24**, 1979 (1964).

—, G. D. BIRNIE, J. BOOHAR, and D. WENTLAND: Fluorinated pyrimidines. XX. Inhibition of the nucleoside phosphorylase cleavage of 5-fluoro-2'-deoxyuridine by 5-trifluoromethyl-2'-deoxyuridine. Biochim. biophys. Acta (Amst.) **76**, 315 (1963).

—, J. BOOHAR, and B. KAMPSCHROER: Fluorinated pyrimidines. XXIV. *In vivo* metabolism of 5-trifluoromethyluracil-2-C^{14} and 5-trifluoromethyl-2'-deoxyuridine-2-C^{14}. Cancer Res. **25**, 377 (1965).

—, N. K. CHAUDHURI, P. DANNEBERG, D. MOOREN, L. GRIESBACH, R. DUSCHINSKY, R. J. SCHNITZER, E. PLEVEN, and J. SCHEINER: Fluorinated pyrimidines, a new class of tumor-inhibitory compounds. Nature **179**, 663 (1957).

—, L. GRIESBACH, B. J. MONTAG, D. MOOREN, O. CRUZ, R. J. SCHNITZER, and E. GRUNBERG: Studies on fluorinated pyrimidines. II. Effects on transplanted tumors. Cancer Res. **18**, 305 (1958).

—, D. G. PARSONS, and D. C. REMY: Synthesis of 5-trifluoromethyluracil and 5-trifluoro-methyl-2'-deoxyuridine. J. Amer. chem. Soc. **84**, 3597 (1962).

— — — Synthesis of 5-trifluoromethyluracil and 5-trifluoromethyl-2'-deoxyuridine. J. med. Chem. **7**, 1 (1964).

HENDERSON, E. S., and P. J. BURKE: Clinical experience with cytosine arabinoside. Cancer Res. **6**, 26 (1965).

HENDERSON, J. F.: Feedback inhibition of purine biosynthesis in ascites tumor cells by purine analogues. Biochem. Pharmacol. **12**, 551 (1963).

— Effects of anticancer drugs on biochemical control mechanisms. Progr. exp. Tumor Res. (Basel) **6**, 85 (1965).

—, I. C. CALDWELL, and A. R. P. PATERSON: Decreased feedback inhibition in a 6-(methyl-mercapto) purine ribonucleoside-resistant tumor. Cancer Res. **27**, 1773 (1967).

—, A. R. P. PATERSON, I. C. CALDWELL, and M. HORI: Biochemical effects of formycin, an adenosine analog. Cancer Res. **27**, 715 (1967).

HIGNETT, R. C.: Interference of 5-fluorouracil in the biosynthesis of ribosomes in *Staphylococcus aureus* (strain Duncan). Biochim. biophys. Acta (Amst.) **114**, 559 (1966).

HILLS, D. C., and J. HOROWITZ: Ribosome synthesis in *Escherichia coli* treated with 5-fluorouracil. Biochemistry **5**, 1625 (1966).

HIRSCHBERG, E., J. KREAM, and A. GELLHORN: Enzymatic deamination of 8-azaguanine in normal and neoplastic tissues. Cancer Res. **12**, 524 (1952).

HITCHINGS, G. H.: Antimetabolites and chemotherapy: integration of biochemistry and molecular manipulation. In: Chemotherapy of Cancer. Ed.: P. A. PLATTNER. New York: Elsevier 1964, p. 77.

— Role of inhibitors in studies of comparative enzymology. Fed. Proc. **26**, 1078 (1967).

—, and J. J. BURCHALL: Inhibition of folate biosynthesis and function as a basis for chemotherapy. In: Advances in Enzymology, Vol. 27. Ed.: F. F. NORD. New York: Interscience 1965, p. 418.

—, and G. B. ELION: Chemical suppression of the immune response. Pharmacol. Rev. **15**, 365 (1963).

— — Mechanisms of action of purine and pyrimidine analogs. In: Cancer Chemotherapy. Eds.: I. BRODSKY, S. B. KAHN, and J. H. MOYER. New York: Grune & Stratton 1967, p. 26.

— —, E. A. FALCO, P. B. RUSSELL, M. B. SHERWOOD, and VANDERWERFF: Antagonists of nucleic acid derivatives. I. The *lactobacillus casei* model. J. biol. Chem. **183**, 1 (1950 a).

HITCHINGS, G. H., G. B. ELION, E. A. FALCO, and H. VANDERWERFF: Studies on analogs of purines and pyrimidines. Ann. N. Y. Acad. Sci. 52, 1318 (1950).

—, E. A. FALCO, H. VANDERWERFF, P. B. RUSSELL, and G. B. ELION: Antagonists of nucleic acid derivatives. VII. 2,4-diaminopyrimidines. J. biol. Chem. 199, 43 (1952).

HO, D. H. W., and E. FREI, III: Biochemical and pharmacological studies of 6-methylthio-purine ribonucleoside. Fed. Proc. 27, 759 (1968).

HOLOUBEK, V.: The composition of tobacco mosaic virus protein after the incorporation of 5-fluorouracil into the virus. J. molec. Biol. 6, 164 (1963).

HORI, M., E. ITO, T. TAKITA, G. KOYOMA, T. TAKEUCHI, and H. UMEZAWA: A new antibiotic, formycin. J. Antibiot. (Tokyo), Ser. A, 17, 96 (1964).

—, T. WAKASHIRO, E. ITO, T. SAWA, T. TAKEUCHI, and H. UMEZAWA: Biochemical effects of formycin B on *Xanthomonas oryzae*. J. Antibiot. (Tokyo) 21, 264 (1968).

HOROWITZ, J., and V. KOHLMEIER: Formation of active β-galactosidase by *E. coli* treated with 5-fluorouracil. Biochim. biophys. Acta (Amst.) 142, 208 (1967).

—, J. J. SAUKKONEN, and E. CHARGAFF: Effect of 5-fluorouracil on a uracil-requiring mutant of *Escherichia coli*. Biochim. biophys. Acta (Amst.) 29, 222 (1959).

— — — Effects of fluoropyrimidines on the synthesis of bacterial protein and nucleic acids. J. biol. Chem. 235, 3266 (1960).

HOWARD, J. P., N. CEVIK, and M. L. MURPHY: Cytosine arabinoside (NSC-63878) in acute leukemia in children. Cancer Chemother. Rept. 50, 287 (1966).

HUBERT-HABART, M., and S. S. COHEN: The toxicity of 9-β-D-arabinofuranosyladenine to purine requiring *Escherichia coli*. Biochim. biophys. Acta (Amst.) 59, 468 (1962).

HURWITZ, J., and J. T. AUGUST: The role of DNA in RNA synthesis. Progr. Nucl. Acid Res. 1, 59 (1963).

HUTCHISON, D. J.: Studies on cross resistance and collateral sensitivity. Cancer Res. 25, 1581 (1965).

IBALL, J., C. H. MORGAN, and H. R. WILSON: Structure of 5-bromodeoxyuridine and 5-bromo-uridine. Nature 209, 1230 (1966).

IKEHARA, M., and T. FUKUI: Some physical properties of poly 7-deazaadenylic acid (poly-tubercidin-phosphoric acid). J. molec. Biol. 38, 437 (1968).

—, K. MURAO, F. HARADA, and S. NISHIMURA: Synthesis of formycin triphosphate and its incorporation into ribopolynucleotide by DNA-dependent RNA polymerase. Biochim. biophys. Acta (Amst.) 155, 82 (1968).

— — — — Protein synthesis directed by ribopolynucleotide containing formycin. Biochim. biophys. Acta (Amst.) 174, 696 (1969).

—, and E. OHTSUKA: Stimulation of the binding of aminoacyl-sRNA to ribosomes by tuber-cidin (7-deazaadenosine) and N^6-dimethyladenosine containing trinucleoside diphosphate analogs. Biochem. biophys. Res. Commun. 21, 257 (1965).

INGRAM, V. M.: The biosynthesis of macromolecules. New York: W. A. Benjamin Inc. 1965.

ISHIZUKA, M., T. SAWA, S. HORI, H. TAKEYAMA, T. TAKEUCHI, and H. UMEZAWA: Biological studies on formycin and formycin B. J. Antibiot. (Tokyo) 21, 5 (1968).

— —, G. KOYOMA, T. TAKEUCHI, and H. UMEZAWA: Metabolism of formycin and formycin B *in vivo*. J. Antibiot. (Tokyo) 21, 1 (1968 a).

JAFFE, J. J., R. E. HANDSCHUMACHER, and A. D. WELCH: Studies on the carcinostatic activity in mice of 6-azauracil riboside (azauridine), in comparison with that of 6-azauracil. Yale J. Biol. Med. 30, 168 (1957).

JAGGER, D. V., N. M. KREDICH, and A. J. GUARINO: Inhibition of Ehrlich mouse ascites tumor growth by cordycepin. Cancer Res. 21, 216 (1961).

JUROVCIK, M., K. RASKA, JR., Z. SORMOVA, and F. SORM: Anabolic transformations of the new antimetabolite 5-azacytidine and its incorporation into ribonucleic acid. Coll. Czech. Chem. Commun. 30, 3370 (1965).

KABAT, S., and D. W. VISSER: The incorporation of aminodeoxyuridine into deoxyribonucleic acid of *Escherichia coli* 15T⁻. Biochim. biophys. Acta (Amst.) 82, 680 (1964).

KACZKA, E. A., E. L. DULANEY, C. O. GITTERMAN, H. B. WOODRUFF, and K. FOLKERS: Isola-tion and inhibitory effects on KB cell cultures of 3'-deoxyadenosine from *Aspergillus nidulans* (Eidam) Wint. Biochem. biophys. Res. Commun. 14, 452 (1964).

KACZKA, E. A., N. R. TRENNER, B. ARISON, R. W. WALKER, and K. FOLKERS: Identification of cordycepin, a metabolite of *Cordyceps militaris,* as 3'-deoxyadenosine. Biochem. biophys. Res. Commun. **14**, 456 (1964 a).

KAHAN, F. M., and J. HURWITZ: The role of deoxyribonucleic acid in ribonucleic acid synthesis. IV. The incorporation of pyrimidine and purine analogues into ribonucleic acid. J. biol. Chem. **237**, 3778 (1962).

KALLE, G. P., and J. S. GOTS: Alterations in purine nucleotide pyrophosphorylases and resistance to purine analogues. Biochim. biophys. Acta (Amst.) **53**, 166 (1961).

KALOUSEK, F., I. RYCHLIK, and F. SORM: Inhibition of formation of the acceptor sequence pCpCpA in soluble ribonucleic acid by 6-azauridine-5'-diphosphate. Biochim. Biophys. Acta (Amst.) **61**, 368 (1962).

KANO, H., Y. NAKAGAWA, H. KOYAMA, and Y. TSUKUDA: Structure of a new class of C-nucleoside antibiotic, showdomycin. Abstracts of the 1st intern. Congr. of Heterocyclic Chem., Albuquerque, New Mexiko, 1967.

KAPLAN, A. S., and T. BEN-PORAT: Mode of antiviral action of 5-iodouracil deoxyriboside. J. molec. Biol. **19**, 320 (1966).

— — Differential incorporation of iododeoxyuridine into the DNA of pseudorabies virus-infected and noninfected cells. Virology **31**, 734 (1967).

— —, and T. KAMIYA: Incorporation of 5-bromodeoxyuridine and 5-iododeoxyuridine into viral DNA and its effect on the infective process. Ann. N. Y. Acad. Sci. **130**, 226 (1965).

—, M. BROWN, and T. BEN-PORAT: Effect of 1-β-D-arabinofuranosylcytosine on DNA synthesis. I. In normal rabbit kidney cell cultures. Molec. Pharmacol. **4**, 131 (1968).

KAPLAN, S., J. NORTHUP, R. C. DECONTI, and P. CALABRESI: Suppression of immunologic responses by cytosine arabinoside. Blood **28**, 1001 (1966).

KARON, M., S. WEISSMAN, C. MEYER, and P. HENRY: Studies of DNA, RNA, and protein synthesis in cultured human cells exposed to 8-azaguanine. Cancer Res. **25**, 185 (1965).

KAUFMAN, H. E.: Clinical cure of herpes simplex keratitis by 5-iodo-2'-deoxyuridine. Proc. Soc. exp. Biol. (N. Y.) **109**, 251 (1962).

— *In vivo* studies with antiviral agents. Ann. N. Y. Acad. Sci. **130**, 168 (1965).

—, J. A. CAPELLA, E. D. MALONEY, J. E. ROBBINS, G. M. COOPER, and M. H. UOTILA: Corneal toxicity of cytosine arabinoside. Arch. Ophthal. **72**, 535 (1964).

—, E. MARTOLA, and C. DOHLMAN: Use of 5-iodo-2'-deoxyuridine (IDU) in treatment of herpes simplex keratitis. Arch. Ophthal. **68**, 235 (1962).

—, A. B. NESBURN, and E. D. MALONEY: IDU therapy of herpes simplex. Arch. Opthal. **67**, 583 (1962).

KAWASHIMA, M.: Nucleic acid antagonists. III. Growth inhibitory effect of 5-phenylazo-pyrimidine derivatives on *Streptococcus faecalis.* Chem. pharm. Bull. Tokyo, **7**, 13 (1959).

KELLEY, W. N., F. M. ROSENBLOOM, J. F. HENDERSON, and J. E. SEEGMILLER: A specific enzyme defect in gout associated with over production of uric acid. Proc. nat. Acad. Sci (Wash.) **57**, 1735 (1967).

KESSEL, D.: Some observations on the phosphorylation of cytosine arabinoside. Molec. Pharmacol. **4**, 402 (1968).

—, T. C. HALL, and I. WOODINSKY: Transport and phosphorylation as factors in the antitumor action of cytosine arabinoside. Science **156**, 1240 (1967).

KIDDER, G. W., and V. C. DEWEY: The biological activity of substituted pyrimidines. J. biol. Chem. **178**, 383 (1949).

KIM, J. H., and M. L. EIDINOFF: Action of 1-β-D-arabinofuranosylcytosine on the nucleic acid metabolism and viability of HeLa cells. Cancer Res. **25**, 698 (1965).

—, A. S. GELBARD, A. G. PEREZ, and M. L. EIDINOFF: Effect of 5-bromodeoxyuridine on nucleic acid and protein synthesis and viability in HeLa cells. Biochim. biophys. Acta (Amst.) **134**, 388 (1967).

KIMBALL, A. P., B. BOWMAN, P. S. BUSH, J. HERRIOTT, and G. A. LEPAGE: Inhibitory effects of the arabinosides of 6-mercaptopurine and cytosine on purine and pyrimidine metabolism. Cancer Res. **26**, 1337 (1966).

KLENOW, H.: Formation of the mono-, di- and triphosphate of cordycepin in Ehrlich ascites-tumor cells *in vitro.* Biochim. biophys. Acta (Amst.) **76**, 347 (1963).

KLENOW, H., and S. FREDERIKSEN: Effect of 3'-deoxyATP (cordycepin triphosphate) and 2'-deoxyATP on the DNA-dependent RNA nucleotidyltransferase from Ehrlich ascites tumor cells. Biochim. biophys. Acta (Amst.) 87, 495 (1964).

KLOET, S. R. DE, and P. J. STRIJKERT: Selective inhibition of ribosomal RNA synthesis in *Saccharomyces carlsbergensis* by 5-fluorouracil. Biochem. biophys. Res. Commun. 23, 49 (1966).

KORNBERG, A.: Enzymatic Synthesis of DNA. New York: John Wiley and Sons 1961.

— Active center of DNA polymerase. Science 163, 1410 (1969).

KOSHIURA, R., and G. A. LePAGE: Some inhibitors of deamination of 9-β-D-arabinofuranosyladenine and 9-β-D-xylofuranosyladenine by blood and neoplasms of experimental animals and humans. Cancer Res. 28, 1014 (1968).

KOYOMA, G., K. MAEDA, H. UMEZAWA, and Y. IITAKA: The structural studies of formycin and formycin B. Tetrahedron Letters 6, 597 (1966).

—, and H. UMEZAWA: Formycin B and its relation to formycin. J. Antibiot. (Tokyo), Ser. A, 18, 175 (1965).

KRAKOFF, I. H.: Chemotherapy and hormonal therapy of carcinoma of the breast. In: Cancer Chemotherapy. Eds.: I. BRODSKY, S. B. KAHN, and J. H. MOYER. New York: Grune & Stratton 1967, p. 77.

KRENITSKY, T. A., G. B. ELION, R. A. STRELITZ, and G. H. HITCHINGS: Ribonucleosides of allopurinol and oxoallopurinol. Isolation from human urine, enzymatic synthesis and characterization. J. biol. Chem. 242, 2675 (1967).

KWAN, S. W., and T. E. WEBB: A study of the mechanism of polyribosome breakdown induced in regenerating liver by 8-azaguanine. J. biol. Chem. 242, 5542 (1967).

LARK, C., and K. G. LARK: Evidence for two distinct aspects of the mechanism regulating chromosome replication in *Escherichia coli*. J. molec. Biol. 10, 120 (1964).

LARSSON, A., and P. REICHARD: Enzymatic reduction of ribonucleotides. Prog. Nucl. Acid Res. molec. Biol. 7, 303 (1967).

LASNITZKI, I., R. E. F. MATHEWS, and J. D. SMITH: Incorporation of 8-azaguanine into nucleic acids. Nature 173, 346 (1954).

LEE, W. W., A. BENITEZ, L. GOODMAN, and B. R. BAKER: Potential anticancer agents. XL. Synthesis of the β-anomer of 9-(D-arabinofuranosyl)-adenine. J. Amer. chem. Soc. 82, 2648 (1960).

LENGYEL, P.: Problems in protein biosynthesis. In: New York Heart Assocn. (spon.), Proc. of a Symp. on Macromolecular Metabolism. Boston: Little Brown & Co. 1966, p. 305.

LePAGE, G. A.: Incorporation of 6-thioguanine into nucleic acids. Cancer Res. 20, 403 (1960).

—, and M. JONES: Further studies on the mechanism of action of 6-thioguanine. Cancer Res. 21, 1590 (1961).

—, I. G. JUNGA: Metabolism of purine nucleoside analogs. Cancer Res. 25, 46 (1965).

— —, and B. BOWMAN: Biochemical and carcinostatic effects of 2'-deoxythioguanosine. Cancer Res. 24, 835 (1964).

LESLIE, J.: Spectral shift in the reaction of N-ethylmaleimide with proteins. Ann. Biochem. 10, 162 (1965).

LEVENE, P. A., and F. B. LaFORGE: Structure of pyrimidine nucleosides. Ber. 45, 616 (1912).

LEVIN, D. H.: The incorporation of 8-azaguanine into soluble ribonucleic acid of *Bacillus cereus*. J. biol. Chem. 238, 1098 (1963).

—, and M. LITT: Evidence for an altered secondary structure of soluble ribonucleic acid containing 8-azaguanine. Abstracts, 6th Intern. Biochem. Congr. 4, 69 (1964).

LEVITT, J., and Y. BECKER: The effect of cytosine arabinoside on the replication of herpes simplex virus. Virology 31, 129 (1967).

LINDBERG, B., H. KLENOW, and K. HANSEN: Some properties of partially purified mammalian adenosine kinase. J. biol. Chem. 242, 350 (1967).

LOMAX, M. I. S., and G. R. GREENBERG: An exchange between the hydrogen atom on carbon 5 of deoxyuridylate and water catalyzed by thymidylate synthetase. J. biol. Chem. 242, 1302 (1967).

LORENSON, M. Y., G. F. MALEY, and F. MALEY: The purification and properties of thymidylate synthetase from chick embryo extracts. J. biol. Chem. 242, 3332 (1967).

LOWRIE, R. J., and P. L. BERGQUIST: Transfer ribonucleic acids from *Escherichia coli* treated with 5-fluorouracil. Biochemistry 7, 1761 (1968).

Lozeron, H. A., and W. Szybalski: Incorporation of 5-fluorodeoxyuridine into the DNA of *Bacillus subtilis* phage PBS2 and its radiobiological consequences. J. molec. Biol. 30, 277 (1967).

Lucas-Lenard, J., and S. S. Cohen: The inhibitory effect of substrate analogues on polynucleotide phosphorylase. Biochim. biophys. Acta (Amst.) 123, 471 (1966).

Mandel, H. G., P. E. Carlo, and P. K. Smith: The incorporation of 8-azaguanine into nucleic acids of tumor-bearing mice. J. biol. Chem. 206, 181 (1954).

—, and R. Markham: The effect of 8-azaguanine on the biosynthesis of ribonucleic acid in *Bacillus cereus.* Biochem. J. 69, 297 (1958).

Mathews, C. K., and S. S. Cohen: Inhibition of phage-induced thymidylate synthetase by 5-fluorodeoxyuridylate. J. biol. Chem. 238, 367 (1963).

Matsuura, S., O. Shiratori, and K. Katagiri: Antitumor activity of showdomycin. J. Antibiot. (Tokyo), Ser. A, 17, 234 (1964).

McCollister, R. J., W. R. Gilbert Jr., D. M. Ashton, and J. B. Wyngaarden: Pseudo-feedback inhibition of purine synthesis by 6-mercaptopurine ribonucleotide and other purine analogues. J. biol. Chem. 239, 1560 (1964).

Meich, R. P., and R. E. Parks Jr.: Inhibition of a nucleoside monophosphate kinase by 6-thioguanosine 5'-monophosphate. Proc. Amer. Ass. Cancer Res. 5, 44 (1964).

— —, J. H. Anderson Jr., and A. C. Sartorelli: An hypothesis on the mechanism of action of 6-thioguanine. Biochem. Pharmacol. 16, 2222 (1967).

Mitchell, J. H., Jr., H. E. Skipper, and L. L. Bennett Jr.: Investigation of the nucleic acids of viscera and tumor tissue from animals injected with radioactive 8-azaguanine. Cancer Res. 10, 647 (1950).

Mitra, S., and A. Kornberg: Enzymatic mechanisms of DNA replication. In: New York Heart Assocn. (spon.), Proc. of a Symp. on Macromolecular Metabolism. Boston: Little Brown & Co. 1966, p. 59.

Mizuno, Y., M. Ikehara, K. A. Watanabe, S. Suzaki, and T. Itoh: Synthetic studies of potential antimetabolites. IX. The anomeric configuration of tubercidin. J. Org. Chem. 28, 3329 (1963).

Modest, E. J., H. N. Schlein, and G. F. Foley: Antimetabolite activity of 5-arylazo-pyrimidines. J. pharm. Pharmacol. 9, 68 (1957).

Momparler, R. L.: Effect of cytosine arbinoside 5'-triphosphate on mammalian DNA polymerase. Biochem. biophys. Res. Commun. 34, 465 (1969).

—, M. Y. Chu, and G. A. Fischer: Studies on a new mechanism of resistance of L5178Y murine leukemia cells to cytosine arabinoside. Biochim. biophys. Acta (Amst.) 161, 481 (1968).

—, and G. A. Fischer: Mammalian deoxynucleoside kinases. I. Deoxycytidine kinase: purification, properties, and kinetic studies with cytosine arabinoside. J. biol. Chem. 243, 4298 (1968).

Montgomery, J. A.: On the chemotherapy of cancer. Progr. Drug Res. 8, 431 (1965).

Moore, E. C., and S. S. Cohen: Effects of arabinonucleotides on ribonucleotide reduction by an enzyme system from rat tumor. J. biol. Chem. 242, 2116 (1967).

—, and G. A. LePage: The metabolism of 6-thioguanine in normal and neoplastic tissues. Cancer Res. 18, 1075 (1958).

Morrell, S. A., V. E. Ayers, T. J. Greenwalt, and P. Hoffman: Thiols of the erythrocytes. Reaction of N-ethylmaleimide with intact erythrocytes. J. biol. Chem. 239, 2696 (1964).

Mosteller, R., J. M. Ravel, and B. Hardesty: Differential inactivation of soluble reticulocyte transfer factors with N-ethylmaleimide. Biochem. biophys. Res. Commun. 24, 714 (1966).

Morton, R. K.: Enzymic synthesis of coenzyme I in relation to chemical control of cell growth. Nature 181, 540 (1958).

— New concepts of the biochemistry of the cell nucleus. Aust. J. biol. Sci. 24, 260 (1961).

Munyon, W., and N. P. Salzman: The incorporation of 5-fluorouracil into poliovirus. Virology 18, 95 (1962).

Nakada, D., and B. Magasanik: The roles of inducer and catabolite repressor in the synthesis of β-galactosidase by *Escherichia coli.* J. molec. Biol. 8, 105 (1964).

Nakamura, G.: Studies on antibiotic actinomycetes. II. J. Antibiot. (Tokyo), Ser. A, 14, 90 (1961).

NAONO, S., and F. GROS: Synthése par *E. coli* d'une phosphatase modifiée en présence d'un analogue pyrimidique. C. R. Acad. Sci. (Paris) 250, 3889 (1960).

NEUFELD, E. F., and C. W. HALL: Inhibition of UDP-D-glucose dehydrogenase by UDP-D-xylose: a possible regulatory mechanism. Biochem. biophys. Res. Commun. 4, 456 (1965).

NICHOLS, W. W.: *In vitro* chromosome breakage induced by arabinosyladenine in human leukocytes. Cancer Res. 24, 1502 (1964).

NISHIMURA, H.: French Patent, M2751 September 21, 1964. Showdomycin, extraction and properties. CA (N. Y.) 62, 2675 b (1965).

—, K. KATAGIRI, K. SATO, M. MAYAMA, and N. SHIMAOKA: Toyocamycin, a new anti-candida antibiotic. J. Antibiotics (Tokyo), Ser. A, 9, 60 (1956).

—, and Y. KOMATSU: Reversal of inhibiting action of showdomycin on the proliferation of *Escherichia coli* by nucleosides and thiol compounds. J. Antibiotics (Tokyo) 21, 250 (1968).

—, M. MAYAMA, Y. KOMATSU, H. KATO, N. SHIMAOKA, and Y. TANAKA: Showdomycin, a new antibiotic from a Streptomyces sp. J. Antibiotics (Tokyo), Ser. A, 17, 148 (1964).

NISHIMURA, S., F. HARADA, and M. IKEHARA: The selective utilization of tubercidin triphosphate as an ATP analogue in DNA-dependent RNA polymerase system. Biochim. biophys. Acta (Amst.) 129, 301 (1966).

OHKUMA, K.: Chemical structure of toyocamycin. J. Antibiotics (Tokyo), Ser. A, 14, 343 (1961).

OSAWA, S.: Biosynthesis of ribosomes in bacterial cells. Progr. Nucl. Acid. Res. molec. Biol. 4, 161 (1965).

OVERGAARD-HANSEN, K.: The inhibition of 5-phosphoribosyl-1-pyrophosphate formation by cordycepin triphosphate in extracts of Ehrlich ascites tumor cells. Biochim. biophys. Acta (Amst.) 80, 504 (1964).

OWEN, S. P., and C. G. SMITH: Cytotoxicity and antitumor properties of the abnormal nucleoside tubercidin. Cancer Chemother. Rept. 36, 19 (1964).

PACES, V., J. DOSKOCIL, and F. SORM: Incorporation of 5-azacytidine into nucleic acids of *Escherichia coli*. Biochim. biophys. Acta (Amst.) 161, 352 (1968).

PAPAC, R., W. A. CREASEY, P. CALABRESI, and A. D. WELCH: Clinical and pharmacological studies with 1-β-arabinofuranosylcytosine (cytosine arabinoside). Proc. Amer. Ass. Cancer Res. 6, 50 (1965).

PARKS, R. E., JR., J. L. WAY, and J. L. DAHL: Nucleotides of fluorinated pyrimidines. Proc. Amer. Ass. Cancer Res. 2, 333 (1958).

PASTERNAK, C. A., and R. E. HANDSCHUMACHER: The biochemical activity of 6-azauridine: Interference with pyrimidine metabolism in transplantable mouse tumors. J. biol. Chem. 234, 2992 (1959).

PATERSON, A. R. P., and M. C. WANG: Stimulation of 6-mercaptopurine anabolism in tumor cells by prior treatment with 6-(methylmercapto) purine ribonucleoside. Fed. Proc. 27, 759 (1968).

PEARSON, H. E., D. L. LAGERBORG, and D. W. VISSER: Chemical inhibitions of Theiler's virus. Proc. Soc. exp. Biol. (N. Y.) 93, 61 (1956).

PINSKY, L., and R. S. KROOTH: Studies on the control of pyrimidine biosynthesis in human diploid cell strains. I. Effect of 6-azauridine on cellular phenotype. Proc. nat. Acad. Sci. (Wash.) 57, 925 (1967).

— — Studies on the control of pyrimidine biosynthesis in human diploid cell strains. II. Effect of 5-azaorotic acid, barbituric acid, and pyrimidine precursors on cellular phenotype. Proc. nat. Acad. Sci. (Wash.) 57, 1267 (1967 a).

PISKALA, A., and F. SORM: Nucleic acid components and their analogues. LI. Synthesis of 1-glycosyl derivatives of 5-azauracil and 5-azacytosine. Coll. Czech. Chem. Commun. 29, 2060 (1964).

PITHOVA, P., A. PISKALA, J. PITHA, and F. SORM: Nucleic acid components and their analogs. LXVI. Hydrolysis of 5-acacytidine and its connection with biological activity. Coll. Czech. Chem. Commun. 30, 2801 (1965).

PIZER, L. I., and S. S. COHEN: Metabolism of pyrimidine arabinonucleotides and cyclo-nucleosides in *Escherichia coli*. J. biol. Chem. 235, 2387 (1960).

PRUSOFF, W. H.: Synthesis and biological activities of iododeoxyuridine, an analog of thymidine. Biochim. biophys. Acta (Amst.) 32, 295 (1959).

PRUSOFF, W. H.: A review of some aspects of 5-iododeoxyuridine and azauridine. Cancer Res. 23, 1246 (1963).
— Recent advances in chemotherapy of viral diseases. Pharmacol. Rev. 19, 209 (1967).
—, Y. S. BAKHLE, and L. SEKELY: Cellular and antiviral effects of halogenated deoxyribonucleosides. Ann. N. Y. Acad. Sci. 130, 135 (1965).
—, and P. K. CHANG: 5-iodo-2′-deoxyuridine 5′-triphosphate, an allosteric inhibitor of deoxycytidylate deaminase. J. biol. Chem. 243, 223 (1968).
—, J. J. JAFFE, and H. GUNTHER: Studies in the mouse of the pharmacology of 5-iododeoxyuridine, an analogue of thymidine. Biochem. Pharmacol. 3, 110 (1960).
PUGH, L. H., and N. N. GERBER: The effect of 3′-amino-3′-deoxyadenosine against ascitic tumors of mice. Cancer Res. 23, 640 (1963).
—, H. A. LECHEVALIER, and M. SOLOTOROVSKY: Antitumor activity of a substance produced by a strain of *Helminthosporium*. Antibiot. and Chemother. 12, 310 (1962).
RAO, K. V.: Abstract, 150th meeting, Amer. chem. Soc., 1965, p. 24.
—, and D. W. RENN: BA-90912: An antitumor substance. Antimicrobial Agents and Chemotherapy, 1963, p. 77.
RASKA, K., JR., M. JUROVCIK, V. FUCIK, R. TYKVA, Z. SORMOVA, and F. SORM: Metabolic effects of 5-azacytidine in isolated nuclei of calf thymus cells. Coll. Czech. Chem. Commun. 31, 2809 (1966).
— —, Z. SORMOVA, and F. SORM: Inhibition by 5-azacytidine of ribonucleic acid synthesis in isolated nuclei of calf thymus. Coll. Czech. Chem. Commun. 30, 3215 (1965).
REGELSON, W.: Chemotherapy of gastrointestinal cancer. In: Cancer Chemotherapy. Eds.: I. BRODSKY, S. B. KAHN, and J. H. MOYER. New York: Grune & Stratton 1967, p. 84.
REICHARD, P.: The enzymic synthesis of pyrimidines. Adv. Enzymol. 21, 263 (1959).
RENIS, H. E., H. G. JOHNSON, and B. K. BHUYAN: A collagen plate assay for cytotoxic agents. I. Methods. Cancer Res. 22, 1126 (1962).
REYES, P., and C. HEIDELBERGER: Fluorinated pyrimidines XXVI. Mammalian thymidylate synthetase: Its mechanism of action and inhibition by fluorinated nucleotides. Molec. Pharmacol. 1, 14 (1965).
RICHARDSON, C. C., C. L. SCHILDKRAUT, and A. KORNBERG: Studies on the replication of DNA by DNA polymerases. Cold Spring Harb. Symp. Quant. Biol. 28, 9 (1963).
ROBERTS, M., and D. W. VISSER: Uridine and cytidine derivatives. J. Amer. chem. Soc. 74, 668 (1952).
— — Antimetabolite activity of uridine and cytidine derivatives. J. biol. Chem. 194, 695 (1952 a).
ROBINS, R. K.: Potential purine antagonists. I. Synthesis of some 4,6-substituted pyrazolo (3,4-d) pyrimidines. J. Amer. chem. Soc. 78, 784 (1956).
—, L. B. TOWNSEND, F. CASSIDY, J. F. GERSTER, A. F. LEWIS, and R. L. MILLER: Structure of the nucleoside antibiotics formycin, formycin B, and laurusin. J. Heterocyclic Chem. 3, 110 (1966).
ROBLIN, R. O., JR., J. O. LAMPEN, J. P. ENGLISH, Q. P. COLE, and J. R. VAUGHAN: Studies in chemotherapy. VIII. Methionine and purine antagonists and their relation to the sulfonamides. J. Amer. chem. Soc. 67, 290 (1945).
ROGERS, H. J., and H. R. PERKINS: 5-fluorouracil and mucopeptide biosynthesis by *Staphylococcus aureus*. Biochem. J. 77, 448 (1960).
ROSEN, B.: Characteristics of 5-fluorouracil-induced synthesis of alkaline phosphatase. J. molec. Biol. 11, 845 (1965).
ROTTMAN, F., and A. J. GUARINO: Studies on the inhibition of *Bacillus subtilis* growth by cordycepin. Biochim. biophys. Acta (Amst.) 80, 632 (1964).
— — The inhibition of phosphoribosyl-pyrophosphate into RNA by RNA polymerase from *Micrococcus lysodeikticus*. Biochim. biophys. Acta (Amst.) 89, 465 (1964 a).
ROY-BURMAN, P.: UDP-glucose dehydrogenase. Studies with uridine diphosphate xylose, 5-hydroxyuridine diphosphate xylose, and 5,6-dihydrouridine diphosphate xylose. Manuscript, 1969.
—, S. ROY-BURMAN, and D. W. VISSER: Nucleotides and polynucleotides of pyrimidine analogues. Fed. Proc. 24, 483 (1965).
— — — Incorporation of 5,6-dihydrouridine triphosphate into ribonucleic acid by DNA-dependent RNA polymerase. Biochem. biophys. Res. Commun. 20, 291 (1965 a).

Roy-Burman, P., S. Roy-Burman, and D. W. Visser: Utilization of 5,6-dihydrouridine 5'-triphosphate in the reaction catalyzed by *Escherichia coli* RNA polymerase. Biochim. biophys. Acta. (Amst.) 142, 355 (1967).

— — — Uridine diphosphate glucose dehydrogenase. Studies with 5-hydroxyuridine diphosphate glucose and 5,6-dihydrouridine diphosphate glucose. J. biol. Chem. 243, 1692 (1968).

—, and D. Sen: N-(5-Arylazo-4-pyrimidyl)-amino acids as growth inhibitors of *Streptococcus faecalis.* Nature 196, 1316 (1962).

— — Effect of a number of N-pyrimidyl amino acids and some of their 5-arylazo derivatives on the growth of certain microorganisms. Biochem. Pharmacol. 13, 1437 (1964).

Roy-Burman, S., P. Roy-Burman, and D. W. Visser: Inhibition of ribonucleic acid polymerase by 5-hydroxyuridine 5'-triphosphate. J. biol. Chem. 241, 781 (1966).

— — — Showdomycin, a new nucleoside antibiotic. Cancer Res. 28, 1605 (1968 a).

— — — Studies on the effects of triphosphates of 5-aminouridine and 5-hydroxydeoxyuridine on RNA and DNA polymerases. Manuscript, 1969.

Rubin, R. J., J. J. Jaffe, and R. E. Handschumacher: Quantitative differences in the pyrimidine metabolism of *Trypanosoma equiperdum* and mammals as characterized by 6-azauracil and 6-azauridine. Biochem. Pharmacol. 11, 563 (1962).

—, A. Reynard, and R. E. Handschumacher: An analysis of the lack of drug synergism during sequential blockade of *de novo* pyrimidine biosynthesis. Cancer Res. 24, 1002 (1964).

Rundles, R. W., G. B. Ellion, and G. H. Hitchings: Allopurinol in the treatment of gout and secondary hyperuricemia. Bull. rheum. Dis. 16, 400 (1966).

—, J. Laszlo, T. Itoga, J. B. Hobson, and F. E. Garrison Jr.: Clinical and hematologic study of 6-[(1-methyl-4-nitro-5-imidazolyl)thio]purine (B. W. 57-322) and related compounds. Cancer Chemother. Rept. 14, 99 (1961).

—, E. N. Metz, and H. R. Silberman: Allopurinol in the treatment of gout. Ann. intern. Med. 64, 229 (1966).

—, H. R. Silberman, G. H. Hitchings, and G. B. Elion: Effects of xanthine oxidase inhibitor on clinical manifestations and purine metabolism in gout. Ann. intern. Med. 60, 717 (1964).

—, J. B. Wyngaarden, G. H. Hitchings, G. B. Elion, and H. R. Silberman: Effects of xanthine oxidase inhibitor on thiopurine metabolism, hyperuricemia and gout. Trans. Ass. Amer. Phycns 76, 126 (1963).

Saneyoshi, M., R. Tokuzen, and F. Fukuoka: Antitumor activities and structural relationship of tubercidin, toyocamycin, and their derivatives. Gann 56, 219 (1965).

Sartorelli, A. C., and G. A. LePage: Metabolic effects of 6-thioguanine. II. Biosynthesis of nucleic acid purines *in vivo* and *in vitro*. Cancer Res. 18, 1329 (1958).

— —, and E. C. Moore: Metabolic effects of 6-thioguanine. I. Studies on thioguanine-resistant and -sensitive Ehrlich ascites cells. Cancer Res. 18, 1232 (1958).

—, H. F. Upchurch, A. L. Bieber, and B. A. Booth: Some metabolic effects exerted by azaserine and purine analogs *in vivo*. Cancer Res. 24, 1202 (1964).

Saslaw, L. D., G. B. Grindey, I. Kline, and V. S. Waravdekar: Sparing action of uridine on the activity of arabinosylcytosine with normal and leukemic mice. Cancer Res. 28, 11 (1968).

—, R. Tomchick, G. B. Grindey, I. Kline, and V. S. Waravdekar: Sparing action of uridine on the antileukemic activity of cytosine arabinoside. Proc. Amer. Ass. Cancer Res. 7, 62 (1966).

Scannell, J. P., and G. H. Hitchings: Thioguanine in deoxyribonucleic acid from tumors of 6-mercaptopurine-treated mice. Proc. Soc. exp. Biol. (N. Y.) 122, 627 (1966).

Schabel, F. M., Jr., W. R. Laster, and H. E. Skipper: Chemotherapy of leukemia L1210 by 6-mercaptopurine (NSC-755) in combination with 6-methylthiopurine ribonucleoside (NSC-40774). Cancer Chemother. Rept. 51, 111 (1967).

Schindler, R., and A. D. Welch: Ribosidation as a means of activating 6-azauracil as an inhibitor of cell reproduction. Science 125, 548 (1957).

SCHNEBLI, H. P., D. L. HILL, and L. L. BENNETT JR.: Purification and properties of adenosine kinase from human tumor cells of type H.Ep. No. 2. J. biol. Chem. **242**, 1997 (1967).

SCHRECKER, A. W., and M. J. URSHEL: Metabolism of 1-β-D-arabinofuranosylcytosine in leukemia L1210: Studies with intact cells. Cancer Res. **28**, 793 (1968).

SCHWARTZ, R. S., J. STACK, and W. DAMESHEK: Effect of 6-mercaptopurine on antibody production. Proc. Soc. exp. Biol. (N. Y.) **99**, 164 (1958).

SCOTT, J. L., J. V. MARINO, and E. P. GABOR: Human leukocyte metabolism *in vitro*. II. The effect of 6-mercaptopurine on formate-^{14}C incorporation into the nucleic acids of acute leukemic leukocytes. Blood **28**, 683 (1966).

SEEGMILLER, J. E.: Clinical staff conference: Biochemical abnormalities in hereditary diseases. Ann. intern. Med. **57**, 472 (1962).

—, A. I. GRAYZEL, L. LASTER, and L. LIDDLE: Uric acid production in gout. J. clin. Invest. **40**, 1304 (1961).

—, F. M. ROSENBLOOM, and W. N. KELLEY: Enzyme defect associated with a sex-linked human neurological disorder and excessive purine synthesis. Science **155**, 1682 (1967).

SEIBERT, W.: Ber. **80**, 494 (1947).

SHAPIRO, L., and J. T. AUGUST: Replication of RNA viruses. III. Utilization of ribonucleotide analogues in the reaction catalyzed by a RNA virus RNA polymerase. J. molec. Biol. **14**, 214 (1965).

SHARPLESS, N. E., and M. FLAVIN: The reactions of amines and amino acids with maleimides. Structure of the reaction products deduced from infrared and nuclear magnetic resonance spectroscopy. Biochemistry **5**, 2963 (1966).

SHAW, R. K., R. N. SHULMAN, J. D. DAVIDSON, D. P. RALL, and E. FREI, III: Studies with the experimental antitumor agent 4-aminopyrazolo (3,4-d) pyrimidine. Cancer **13**, 482 (1960).

SHEEN, M. R., B. K. KIM, and R. E. PARKS JR.: Purine nucleoside phosphorylase from human erythrocytes. III. Inhibition by the inosine analog formycin B of the isolated enzyme and of nucleoside metabolism in intact erythrocytes and sarcoma 180 cells. Molec. Pharmacol. **4**, 293 (1968).

SHIGEURA, H. T., and G. E. BOXER: Incorporation of 3'-deoxyadenosine-5'-triphosphate into RNA by RNA polymerase from *Micrococcus lysodeikticus*. Biochim. biophys. Res. Common. **17**, 758 (1964).

— —, M. L. MELONI, and S. D. SAMPON: Structure-activity relationship of some purine 3'-deoxyribonucleotides. Biochemistry **5**, 994 (1966).

—, and C. N. GORDON: The effects of 3'-deoxyadenosine on the synthesis of ribonucleic acid. J. biol. Chem. **240**, 806 (1965).

SHIMURA, Y., R. E. MOSES, and D. NATHANS: Coliphage MS2 containing 5-fluorouracil. II. RNA-deficient particles formed in the presence of 5-fluorouracil. J. molec. Biol. **28**, 95 (1967).

SILAGI, S.: Metabolism of 1-β-D-arabinofuranosylcytosine in L cells. Cancer Res. **25**, 1446 (1965).

SILBERMAN, H. R., and J. B. WYNGAARDEN: 6-mercaptopurine as a substrate and inhibitor of xanthine oxidase. Biochim. biophys. Acta (Amst.) **47**, 178 (1961).

SIMON, E. H.: Effects of 5-bromodeoxyuridine on cell division and DNA replication in HeLa cells. Exp. Cell Res. (Suppl.) **9**, 263 (1963).

SKODA, J.: Mechanism of action and application of azapyrimidines. Progr. Nucl. Acid Res. **2**, 197 (1963).

—, V. F. HESS, and F. SORM: The biosynthesis of 6-azauracil riboside by *Escherichia coli* growing in the presence of 6-azauracil. Experientia **13**, 150 (1957).

—, and F. SORM: Accumulation of nucleic acid metabolites in *Escherichia coli* exposed to the action of 6-azauracil. Biochim. Biophys. Acta **28**, 659 (1958).

— — The accumulation of orotic acid, uracil and hypoxanthine by *Escherichia coli* in the presence of 6-azauracil, and the biosynthesis of 6-azauridylic acid. Coll. Czech. Chem. Commun. **24**, 1331 (1959).

SKOLD, O.: Uridine kinase from Ehrlich ascites tumor: Purification and properties. J. biol. Chem. **235**, 3273 (1960).

SLECHTA, L.: Effect of arabinonucleosides on growth and metabolism of *E. coli*. Fed. Proc. **20**, 357 (1961).

SLOTNICK, I. J., D. W. VISSER, and S. C. RITTENBERG: Growth inhibition of purine-requiring mutants of *Escherichia coli* by 5-hydroxyuridine. J. biol. Chem. 203, 647 (1953).

SMELLIE, R. M. S.: The biosynthesis of ribonucleic acid in animal systems. Progr. in Nucl. Acid Res. 1, 27 (1963).

SMITH, C. G., W. L. LUMMIS, and J. E. GRADY: An improved tissue culture assay. II. Cytotoxicity studies with antibiotics, chemicals, and solvents. Cancer Res. 19, 847 (1959).

SMITH, D. A., and D. W. VISSER: Studies on 5-hydroxyuridine. J. biol. Chem. 240, 446 (1965).

—, P. ROY-BURMAN, and D. W. VISSER: Studies on 5-aminouridine. Biochim. biophys. Acta (Amst.) 119, 221 (1966).

SMITH, H. H., C. P. FUSSELL, and B. H. KUGELMAN: Partial synchronization of nuclear divisions in root meristems with 5-aminouracil. Science 142, 595 (1963).

SMITH, J. D., and R. E. F. MATTHEWS: The metabolism of 8-azapurines. Biochem. J. 66, 323 (1957).

SMITH, K. O.: Some biologic aspects of herpes virus-cell interactions in the presence of 5-iodo-2'-deoxyuridine (IDU). J. immunol. 91, 582 (1963).

—, and C. D. DUKES: Effects of 5-iodo-2-desoxyuridine (IDU) on herpesvirus synthesis and survival in infected cells. J. immunol. 92, 550 (1964).

SORM, F., A. JAKUBOVIC, and L. SLECHTA: The anticancerous action of 6-azauracil (3,5-dioxo-2,3,4,5-tetrahydro-1,2,4-triazine). Experientia 12, 271 (1956).

—, and H. KEILOVA: The anti-tumour activity of 6-azauracil riboside. Experientia 14, 215 (1958).

—, A. PISKALA, A. CIHAK, and J. VESELY: 5-Azacytidine, a new, highly effective carcinostatic. Experientia 20, 202 (1964).

—, and J. SKODA: 6-Azauracil, an antimetabolite of uracil and cytosine in *Escherichia coli*. Preliminary communication. Coll. Czech. Chem. Commun. 21, 487 (1956).

—, and J. VESELY: The activity of a new antimetabolite, 5-azacytidine, against lymphoid leukemia in AK mice. Neoplasia 11, 123 (1964).

SPIEGELMAN, S., H. O. HALVORSON, and R. BEN-ISHAI: Free amino acids and the enzyme-forming mechanism. In: A Symposium on Amino Acid Metabolism. Eds.: W. D. MCELROY and B. GLASS. Baltimore: The Johns Hopkins Press 1955, p. 124.

STADTMAN, E. R.: Allosteric regulation of enzyme activity. In: Advances in Enzymology, Vol. 28. Ed.: F. F. NORD. New York: Interscience 1966, p. 41.

STICKGOLD, R. A., and F. C. NEUHAUS: On the initial stage in peptidoglycan synthesis. Effect of 5-fluorouracil substitution on phospho-N-acetylmuramyl-pentapeptide translocase (uridine 5'-phosphate). J. biol. Chem. 242, 1331 (1967).

STUTTS, P., and R. W. BROCKMAN: A biochemical basis for resistance of L1210 mouse leukemia to 6-thioguanine. Biochem. Pharmacol. 12, 97 (1963).

SUEOKA, N.: Mechanisms of replication and repair of nucleic acid. In: Molecular Genetics, Part 2. Ed.: J. H. TAYLOR. New York: Academic Press 1967, p. 1.

SUHADOLNIK, R. J., and J. G. CORY: Further evidence for the biosynthesis of cordycepin and proof of the structure of 3-deoxyribose. Biochim. biophys. Acta (Amst.) 91, 661 (1964).

— — Effect of cordycepin triphosphate (3'-deoxyadenosine-5'-triphosphate) on the *E. coli* DNA polymerase system. Abstract 150th Meeting, Amer. chem. Soc., 1965, p. 86 c.

—, S. I. FINKEL, and B. M. CHASSY: Nucleoside antibiotics. I. Biochemical tools for studying the structural requirements for interaction at the catalytic and regulatory sites of ribonucleotide reductase from *Lactobacillus leichmannii*. J. biol. Chem. 243, 3532 (1968).

—, T. UEMATSU, and H. UEMATSU: Toyocamycin: phosphorylation and incorporation into RNA and DNA and the biochemical properties of the triphosphate. Biochim. biophys. Acta (Amst.) 149, 41 (1967).

— — —, and R. G. WILSON: The incorporation of sangivamycin 5'-triphosphate into polynucleotide by ribonucleic acid polymerase from *Micrococcus lysodeikticus*. J. biol. Chem. 243, 2761 (1968).

SUGINO, Y.: Metabolism of deoxyribonucleotides. Ann. Rept., Inst. Virus Res., Kyoto Univ. 9, 1 (1965).

SUOLINNA, E. M., K. SLAVIK, and M. T. HAKALA: Content of TMP-synthetase (TMP-S) and inhibition by 5-fluorodeoxyuridine (FUdR) in sarcoma 180 cells (S-180). Fed. Proc. 26, 813 (1967).

SUZUKI, S., and S. MARUMO: Chemical Structure of tubercidin. J. Antibiot. (Tokyo), Ser. A, 13, 360 (1960).
— — Chemical structure of tubercidin. J. Antibiot. (Tokyo), Ser. A, 14, 34 (1961).
SZYBALSKI, W., N. K. COHN, and C. HEIDELBERGER: Effects of 5-trifluoromethyl-2'-deoxy-uridine (F_3TdR) on the biochemical and radiobiological properties of human cells. Fed. Proc. 22, 532 (1963).
TAKEUCHI, T., J. IWANAGA, T. AOYAGI, and H. UMEZAWA: Antiviral effect of formycin and formycin B. J. Antibiot. (Tokyo), Ser. A, 19, 286 (1966).
TALLEY, R. W., R. M. O'BRYAN, W. G. TUCKER, and R. V. LOO: Clinical pharmacology and human antitumor activity of cytosine arabinoside. Cancer 20, 809 (1967).
THEIL, E. C., and S. ZAMENHOF: Studies on 6-methylaminopurine (6-methyladenine) in bacterial deoxyribonucleic acid. J. biol. Chem. 238, 3058 (1963).
THIRY, L.: Viruses grown in the presence of base analogs: specific alteration of susceptibility to inactivation by radiations, mutagens, and protodyes. Virology 28, 543 (1966).
TIMMIS, G. M.: Chemotherapy of Cancer: The Antimetabolite Approach. London: Butterworths 1967.
—, D. G. I. FELTON, H. O. J. COLLIER, and P. L. HUSKINSON: Structure-activity relations in two new series of antifolic acids. J. pharm. Pharmacol. 9, 46 (1957).
TOLIVER, A., and E. H. SIMON: DNA synthesis in 5-bromouracil tolerant HeLa cells. An autoradiographic study. Exp. Cell Res. 45, 603 (1967).
TOMASZ, A., and E. BOREK: An early phase in the bactericidal action of 5-fluorouracil on E. coli K_{12}: Osmotic imbalance. Proc. nat. Acad. Sci. (Wash.) 45, 929 (1959).
— — The mechanism of bacterial fragility produced by 5-fluorouracil: The accumulation of cell wall precursors. Proc. nat. Acad. Sci. (Wash.) 46, 324 (1960).
— — The mechanism of an osmotic instability induced in E. coli K-12 by 5-fluorouracil. Biochemistry 1, 543 (1962).
TRUMAN, J. T., and H. KLENOW: Effect of 3'-amino-3'-deoxyadenosine on nucleic acid synthesis in Ehrlich ascites tumor cells. Molec. Pharmacol. 4, 77 (1968).
TSUKADA, I., T. KUNIMOTO, M. HORI, and T. KOMAI: Isolation and properties of the enzyme which catalyzes the oxidation of formycin B. J. Antibiot. (Tokyo) 22, 36 (1969).
UEDA, T.: Studies on coenzyme analogs. III. Synthesis of 5-substituted uridine 5'-phosphates. Chem. pharm. Bull. 8, 455 (1960).
ULBRICHT, T. L. V.: 5-Hydroxymethylpyrimidines and their derivatives. Progr. Nucl. Acid Res. molec. Biol. 4, 189 (1965).
UMEZAWA, H., T. SAWA, Y. FUKAGAWA, I. HOMMA, M. ISHIZUKA, and T. TAKEUCHI: Studies on formycin and formycin B in cells of Ehrlich carcinoma and E. coli. J. Antibiot. (Tokyo), Ser. A, 20, 308 (1967).
UNDERWOOD, G. E., G. A. ELLIOT, and D. A. BUTHALA: Herpes keratitis in rabbits: pathogenesis and effect of antiviral nucleosides. Ann. N. Y. Acad. Sci. 130, 151 (1965).
UNGER, K. W., and R. SILBER: Studies on the formate activating enzyme: kinetics of 6-mercaptopurine inhibition and stabilization of the enzyme. Biochim. biophys. Acta (Amst.) 89, 167 (1964).
URETSKY, S. C., G. ACS, E. REICH, M. MORI, and L. ALTWERGER: Pyrrolopyrimidine nucleotides and protein synthesis. J. biol. Chem. 243, 306 (1968).
UTTER, M. F.: Guanosine and inosine nucleotides. In: The Enzymes, Vol. 2. Eds.: P. D. BOYER, H. LARDY, and K. MYRBACK. New York: Academic Press 1960, p. 78.
VESELY, J., A. CIHAK, and F. SORM: Biochemical mechanisms of drug resistance. IV. Development of resistance to 5-azacytidine and simultaneous depressions of pyrimidine metabolism in leukemic mice. Intern. J. Cancer 2, 639 (1967).
— — — Characteristics of mouse leukemic cells resistant to 5-azacytidine and 5-aza-2'-deoxycytidine. Cancer Res. 28, 1995 (1968).
— — — Enhanced triphosphatase activity of mouse leukemic cells resistant to 5-azacytidine. Coll. Czech. Chem. Commun. 33, 341 (1968 a).
—, J. SEIFERT, A. CIHAK, and F. SORM: Biochemical changes associated with the development of resistance to 5-azacytidine in AKR leukemic mice. Intern. J. Cancer. 1, 31 (1966).
VISSER, D. W.: 5-Hydroxyuridine. In: Synthetic Procedures in Nucleic Acid Chemistry, Vol. 1. Eds.: W. W. ZORBACH and R. S. TIPSON. New York: Interscience 1968, p. 428.

Visser, D. W., D. L. Lagerborg, and H. E. Pearson: Inhibition of mouse encephalomyelitis virus, *in vitro*, by certain nucleoprotein derivatives. Proc. Soc. exp. Biol. (N. Y.) 79, 571 (1952).

—, and P. Roy-Burman: 5-Hydroxyuridine 5'-phosphate derivatives. In: Synthetic Procedures in Nucleic Acid Chemistry, Vol. 1. Eds.: W. W. Zorbach and R. S. Tipson. New York: Interscience 1968, p. 493.

Wacker, A., S. Kirschfeld, D. Hartmann u. D. Weinblum: Über den Einbau von 5-nitrouracil, 5-aminouracil und 2-thiothymin in die Bakterien-Desoxyribonukleinsäure. J. molec. Biol. 2, 69 (1960).

Wainwright, S. D., and L. K. Wainwright: Regulation of the initiation of hemoglobin synthesis in the blood island cells of chick embryos. III. Qualitative studies of the effects of 8-azaguanine and preparations of transfer RNAs. Cann. J. Biochem. 45, 255 (1967).

Walwick, E. R., W. K. Roberts, and C. A. Dekker: Cyclisation during the phosphorylation of uridine and cytidine by polyphosphoric acid. Proc. chem. Soc. 1959, p. 84.

Wang, M. C., A. I. Simpson, and A. R. P. Paterson: Combinations of 6-mercaptopurine (NSC-755) and 6-(methylmercapto) purine ribonucleoside (NSC-40774) in therapy of Ehrlich ascites carcinoma. Cancer Chemother. Rept. 51, 101 (1967).

Ward, D. C., and E. Reich: Conformational properties of polyformycin: A polyribonucleotide with individual residues in the syn conformation. Proc. nat. Acad. Sci. (Wash.) 61, 1494 (1968).

Watson, J. D., and F. H. C. Crick: Genetical implications of the structure of deoxyribonucleic acid. Nature 171, 964 (1953).

Way, J. L., J. L. Dahl, and R. E. Parks Jr.: Polyphosphate nucleosides of purine analogues. J. biol. Chem. 234, 1241 (1959).

—, and R. E. Parks Jr.: Enzymatic synthesis of 5'-phosphate nucleotides of purine analogues. J. biol. Chem. 231, 467 (1958).

Webb, T. E.: Polyribosome breakdown in rat liver following administration of 8-azaguanine. Biochim. biophys. Acta (Amst.) 138, 307 (1967).

Weinstein, I. B., and D. Grunberger: Coding properties of sRNA containing 8-azaguanine. Biochem. biophys. Res. Commun. 19, 647 (1965).

Welch, A. D., R. E. Handschumacher, and J. J. Jaffe: Studies on the pharmacology of 6-azauracil. J. Pharmacol. exp. Therap. 129, 262 (1960).

—, and W. H. Prusoff: A synopsis of recent investigations of 5-iodo-2'-deoxyuridine. Cancer Chemother. Rept. 6, 29 (1960).

Werkheiser, W. C., and D. W. Visser: Metabolic inhibitors and nucleotide turnover. II. Inhibition of uptake of C^{14} precursors in rat liver and hepatoma slices. Cancer Res. 15, 644 (1955).

—, R. J. Winzler, and D. W. Visser: Metabolic inhibitors and nucleotide turnover. I. Inhibition of uptake of radiophosphate in rat liver and hepatoma slices. Cancer Res. 15, 641 (1955).

White, F. R.: 4-Aminopyrazolo (3,4-d) pyrimidine and three derivatives. Cancer Chemother. Rept. 3, 26 (1959).

Willen, R., and U. Stenram: RNA synthesis in the liver of rats treated with 5-fluorouracil. Arch. Biochem. 119, 501 (1967).

Wilmanns, W.: On the mode of action of 6-mercaptopurine. Klin. Wschr. 40, 1170 (1962).

Wodinsky, I., and C. J. Kensler: Activity of cytosine arabinoside (NSC-63878) in a spectrum of rodent tumors. Cancer Chemother. Rept. 47, 65 (1965).

Wyngaarden, J. B.: Gout. In: The Metabolic Basis of Inherited Disease. Eds.: J. B. Stanbury, J. B. Wyngaarden, and D. S. Frederickson. New York: McGraw-Hill 1966. p. 667.

—, R. W. Rundles, H. R. Silberman, and S. Hunter: Control of hyperuricemia with hydroxypyrazolopyrimidine, a purine analog which inhibits uric acid synthesis. Arthr. and Rheum. 6, 306 (1963).

York, J. L., and G. A. LePage: A proposed mechanism for the action of 9-β-D-arabinofuranosyladenine as an inhibitor of the growth of some ascites cells. Can. J. Biochem. 44, 19 (1966).

Young, R. S. K., and G. A. Fischer: The action of arabinosylcytosine on synchronously growing populations of mammalian cells. Biochem. biophys. Res. Commun. 32, 23 (1968).

ZADRAZIL, S., V. FUCIK, P. BARTL, Z. SORMOVA, and F. SORM: The structure of DNA from *Escherichia coli* cultured in the presence of 5-azacytidine. Biochim. biophys. Acta (Amst.) 108, 701 (1965).

ZAMECNIK, M. V., and P. C. ZAMECNIK: Mutation of chloroplasts in Ageratum, following treatment with 5-bromodeoxyuridine. Exp. Cell Res. 45, 218 (1967).

ZARUBA, F., A. KUTA, and J. ELIS: Treatment of mycosis fungoides with 6-azauracilriboside. Lancet 1963 I, p. 275.

ZIMMERMAN, E. F., and S. A. GREENBERG: Inhibition of protein synthesis by 8-azaguanine. I. Effects on polyribosomes in HeLa cells. Molec. Pharmacol. 1, 113 (1965).

—, B. W. HOLLER, and G. D. PEARSON: Azaguanine inhibition of protein synthesis. II. Effects of poly (U) in *Bacillus cereus*. Biochim. biophys. Acta (Amst.) 134, 402 (1967).

Subject Index

Type-setting, printing and binding: Konrad Triltsch, Graphischer Betrieb, 87 Würzburg, Germany

Monographs already Published

In Production

If you have any concerns about our products,
you can contact us on
ProductSafety@springernature.com

In case Publisher is established outside the EU,
the EU authorized representative is:
Springer Nature Customer Service Center GmbH
Europaplatz 3, 69115 Heidelberg, Germany

Printed by Libri Plureos GmbH
in Hamburg, Germany